This wide-ranging series aims to cover all areas of environmental chemistry, placing emphasis on both basic scientific and pollution-orientated aspects. A central core of text books, suitable for those taking courses in Environmental Sciences, Ecology and Chemistry, provides comprehensive coverage at the undergraduate and first-year postgraduate level of atmospheric chemistry, chemical sedimentology, freshwater chemistry, marine chemistry and soil chemistry. At a more advanced level, the series contains topical accounts of current research interest.

RADIOACTIVE AEROSOLS

CAMBRIDGE ENVIRONMENTAL CHEMISTRY SERIES

Series Editors:

P. G. C. Campbell, *Centre for Advanced Analytical Chemistry, CSIRO, Australia*

J. N. Galloway, *Department of Environmental Science, University of Virginia, USA*

R. M. Harrison, *Department of Chemistry, University of Essex, England*

Other books in this series:

1 Peter Brimblecombe *Air Composition & Chemistry*

2 Malcolm Cresser and Anthony Edwards *Acidification of Freshwaters*

RADIOACTIVE AEROSOLS

○ ○

A. C. CHAMBERLAIN

Formerly Head of Aerosol Group, Atomic Energy Research Establishment, Harwell, UK

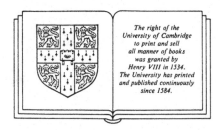

The right of the
University of Cambridge
to print and sell
all manner of books
was granted by
Henry VIII in 1534.
The University has printed
and published continuously
since 1584.

CAMBRIDGE UNIVERSITY PRESS

Cambridge

New York   Port Chester

Melbourne   Sydney

PUBLISHED BY THE PRESS SYNDICATE OF THE UNIVERSITY OF CAMBRIDGE
The Pitt Building, Trumpington Street, Cambridge, United Kingdom

CAMBRIDGE UNIVERSITY PRESS
The Edinburgh Building, Cambridge CB2 2RU, UK
40 West 20th Street, New York NY 10011–4211, USA
477 Williamstown Road, Port Melbourne, VIC 3207, Australia
Ruiz de Alarcón 13, 28014 Madrid, Spain
Dock House, The Waterfront, Cape Town 8001, South Africa

http://www.cambridge.org

First published 1991
First paperback edition 2004

A catalogue record for this book is available from the British Library

Library of Congress cataloguing in publication data
Chamberlain, A. C.
Radioactive aerosols / A.C. Chamberlain.
   p.   cm.   (Cambridge environmental chemistry series)
Includes index.
ISBN 0 521 40121 6 hardback
1. Aerosols, Radioactive – Environmental aspects. 2. Aerosols,
Radioactive – Health aspects.   I. Title.   II. Series.
TD887.R3C43   1991
628.5'35–dc20   90-4316   CIP

ISBN 0 521 40121 6 hardback
ISBN 0 521 61205 5 paperback

CONTENTS

O O O O O O O O O O O O O O O O O O O

PREFACE

○ ○ ○ ○ ○ ○ ○ ○ ○ ○ ○ ○ ○ ○ ○ ○ ○ ○ ○ ○

Public interest in radioactive aerosols began in the mid-1950s, when world-wide fallout of fission products from bomb tests was first observed. The H-bomb test at Bikini Atoll in 1954 had tragic consequences for the Japanese fisherman, and the inhabitants of the Rongelap Atoll, who were in the path of the fallout. In 1957, radio-iodine and other fission products, released in the accident to the Windscale reactor, were tracked over much of Europe, and these events were repeated on a much larger scale after the Chernobyl accident.

Everyone learns from their mistakes, but, in the nuclear industry, it was also the policy from the start to anticipate trouble by calculating the probable consequences of exposure to radioactive materials. Various pathways of exposure had to be considered, including radiation from radioactive clouds and from fallout on the ground, activity inhaled and activity entering via food chains. Only very limited information was available from actual cases of exposure to radioactive aerosols, and this remains the position today. Almost the only epidemiological evidence is related to the exposure of workers in uranium and other mines to radon and its decay products, and much effort has been devoted to understanding the very difficult dosimetric problems which relate to this exposure. Within the last decade it has been realised that domestic exposure to radon, though lower by order of magnitude than that received by miners, is considerably the most important constituent of the population dose, the radiation dose multiplied by the number of persons receiving it.

Over the last 40 years, Harwell Laboratory has contributed in one way or another to the study of radioactive aerosols, both in its theoretical and practical aspects. Also, aerosols have been used experimentally, particularly in the study of the interaction between airborne gases or particles and the surfaces over which they travel. Transfer to surfaces across boundary layers, whether near the ground or in the human lung, is an essential part of the pathway of entry of aerosols into the human body, and the concepts are relevant to other problems.

The first five chapters of this book are about radioactive nuclides of potential concern to public health. In the sixth chapter, some applications to the study of boundary layer transport are discussed. In the last chapter examples are given of using radioactive aerosols to study deposition of particles in the lung and the subsequent uptake into the body. The widespread dissemination of lead aerosol from motor exhausts, its inhalation by the public, and fallout onto crops, present problems of analysis not dissimilar to those of radioactive emissions. Experiments in which volunteers inhaled motor exhaust labelled with ^{203}Pb provided one piece of evidence to fit into the picture.

Some subjects, for example the movement of radioactive particles in the earth's atmosphere and the resulting patterns of fallout are not discussed, being adequately covered in other texts.

No attempt is made to summarise the findings of the International Commission on Radiological Protection, or the reports of the National Radiological Protection Board, except in a few instances where they are directly relevant to the subject under discussion.

There is inevitably a bias towards work done at Harwell. One of the objects of this book is to relate the Harwell work to that done elsewhere and to indicate the recent developments. I am greatly indebted to my colleagues for their help in writing this account.

Many lessons were learnt during the period 1955–1965 by analysis of dispersion and fallout from bomb tests and also from the Windscale accident, but some of these had been forgotten by 1986 when the Chernobyl accident happened, so no apology is made for describing some work which is now 30 years old.

March 1990   A. C. Chamberlain

1

○ ○

Radon

1.1   Physical properties

The gases radon (^{222}Rn) and thoron (^{220}Rn) are formed as progeny of uranium and thorium in rocks and soil. They are emitted from the ground into the atmosphere, where they decay and form daughter products, isotopes of polonium, bismuth and lead, which either remain airborne till they decay, or are deposited in rain and by diffusion to the ground.

Radon and thoron and their decay products are the most important sources of radiation exposure to the general public, contributing on average about half of the total effective dose equivalent received from natural and man-made radioactivity (Clarke & Southwood, 1989).

The emanation of a radioactive gas from radium was observed by Madame Curie. In the atmosphere, radon diffuses and mixes with air like any other gas. Rutherford & Brooks (1901) obtained a value 8×10^{-6} m^2 s^{-1} for the diffusivity of radon. Recent determinations are in the range 1.0 to 1.2×10^{-5} m^2 s^{-1} at N.T.P. (Jost, 1960).

Radon is slightly soluble in water, and obeys Henry's Law. At 20°C the partition coefficient (amount of radon per litre of water at equilibrium divided by the amount per litre of air) is 0.26. Despite the low solubility, water supplies derived locally from granite and metamorphic rocks can be an important source of airborne radon in dwellings (Nero & Nazaroff, 1984; Hess *et al.*, 1987). Radon is more soluble in fats and organic liquids, and the partition coefficient between air and human fat is about unity at 37°C.

The radioactive decay schemes of radon and thoron are shown in Fig. 1.1. The old generic nomenclature (RaA, ThB etc.) is now superseded by the isotopic designation (^{218}Po, ^{212}Pb etc.), but where necessary for clarity the old designation will be added.

Fig. 1. Radioactive decay series.

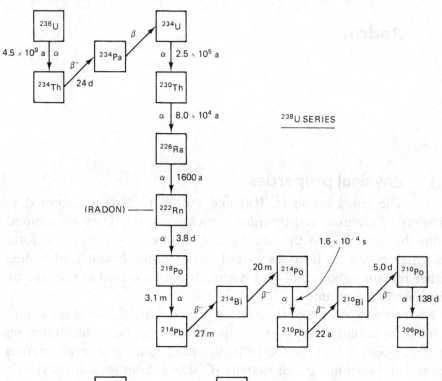

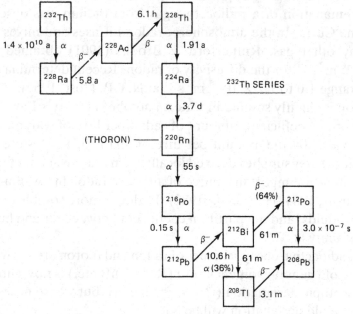

1.2   Radon in air – methods of measurement

Methods of measuring radon in air, have been reviewed by the US National Council on Radiation Protection & Measurement (1988). The most commonly used methods are as follows.

(a) Measurement of activity in ionisation chambers

The air containing the radon is passed into a chamber. After sufficient time for the decay products down to ^{214}Po to reach equilibrium with radon, the activity is assessed from the ionisation current. To allow for radioactive contamination in the materials of the chamber, two identical chambers have been used, one filled with the radon-bearing air, the other with aged air, and with the ionisation currents opposed. Using steel chambers of 6.3 l capacity, Hultqvist (1956) measured 4 Bq m^{-3} of ^{222}Rn with 10% accuracy.

An alternative, for low activities of ^{222}Rn, is to count individual alpha pulses in the ionisation chamber. Kraner et al. (1964) used this method to measure exhalation of ^{222}Rn from the soil. To obtain maximum sensitivity, radon from a large volume of air is adsorbed in activated charcoal, and transferred to an alpha-pulse ionisation chamber in a flow of inert gas.

Ionisation chambers have also been used in a flow-through mode, to give continuous measurement of radon. Israel & Israel (1965) used a very large ionisation chamber, 324 l in volume, for simultaneous measurement of both ^{222}Rn and ^{220}Rn. Air was drawn at 700 l min^{-1} through a filter into the chamber, and was periodically switched through a 2300-l delay vessel, where the ^{220}Rn decayed, before entering the chamber. By using a vibrating reed electrometer to measure the ionisation current, sensitivities of about 0.3 Bq m^{-3} for both ^{222}Rn and ^{224}Rn were obtained. Ionisation chambers are still used for absolute measurements, but for practical purposes they have been superseded by scintillation chambers.

(b) Scintillation chambers

Air with radon is passed into a vessel coated internally with zinc sulphide. Alpha particles from radon in the chamber, and from decay products deposited on the walls, give scintillations which are counted by photomultiplier tubes viewing the chamber through windows (Lucas, 1957). With a chamber of volume 0.1 l, and a counting time of 1 h the detection limit of ^{222}Rn in air was about 10 Bq m^{-3}, but by concentrating the radon in charcoal before measurement Lucas (1964) claimed a sensitivity as low as 4 mBq m^{-3}.

The scintillation chamber can also be used as a flow-through instrument. Because it takes about 2 h for the alpha decay of ^{214}Po to come into equilibrium with the ^{222}Rn in the chamber, calculation is needed to calibrate the instrument. Watnick *et al.* (1986) devised a way round this difficulty by arranging for only the alphas from ^{218}Po (RaA) to be counted. The decay products produced in the chamber were collected electrostatically on the face of a solid state detector, which had resolution sufficient to separate the alpha particles from ^{218}Po from those from ^{214}Po. This enabled a fast response to be obtained, with sensitivity about 40 Bq m^{-3} of ^{222}Rn.

(c) Two filter method
Air is drawn through an efficient filter to remove decay products, and thence through a chamber where fresh decay products are formed. These are collected on an exit filter, and activity is measured either after removal from the chamber, or in situ, usually by alpha scintillation counting. Some decay products are deposited on the walls of the chamber, and their activity can be measured by using a removable lining, or calculated theoretically (Thomas & Le Clare, 1970). If the flow of radon is continued for several hours, the activity of the short-lived products ^{218}Po, ^{214}Pb and ^{214}Bi on the exit filter and walls is the same as the activity of ^{222}Rn in the chamber, and independent of the flow rate through it, so this method can give an absolute calibration.

In applying this method to measurement of ^{222}Rn in air, Schery *et al.* (1980) used a wide chamber (0.76 m diameter, 0.6 m^{-3} volume) to minimise deposition of decay products on the walls. The incoming air was passed through a long pipe in which ^{220}Rn decayed before entering the chamber. With a flow of 200 to 300 l min^{-1}, and counting periods of one or two hours, a sensitivity of 0.4 Bq m^{-3} was obtained.

(d) Passive radon samplers
To meet the need to monitor levels of ^{222}Rn in houses, passive samplers have been developed which measure average concentrations over long periods and do not need power suplies. In the Karlsruhe dosimeter (Urban & Piesch, 1981), a polycarbonate nuclear track detector foil is mounted inside a plastic cup. The mouth of the cup is closed with a filter to allow radon to enter but to exclude decay products. After exposure, the detector foil is etched and the tracks counted optically. This is a

cheap and reliable method, but the sensitivity is low, and an exposure period of months is needed to measure normal indoor concentrations. However, this has the advantage that seasonal fluctuations are evened out.

In a passive detector developed by the National Radiological Protection Board (Wrixon *et al.*, 1988), the etched pits in the detectors are filled with scintillator fluid. After exposure to radon, the detector is irradiated with an alpha source, and the resulting scintillations counted with a photo-multiplier tube. In this way, track density over 1 cm^2 of detector can be measured in a few seconds. Passive detectors used in the UK National Survey were sensitive down to 20 kBq m^{-3} h of accumulated exposure, equivalent to a radon concentration of 5 Bq m^{-3} measured over 4000 h exposure.

George (1984) developed a passive sampler in which radon is adsorbed in a canister containing activated charcoal. After exposure, the canister is returned to the laboratory, and the activity of radon, with decay products in equilibrium, measured by gamma spectrometry. The amount of radon adsorbed was found to depend on the relative humidity, and the calibration depended on the amount of water absorbed, as shown by the gain in weight of the canister. Because radon can desorb from charcoal at ambient temperatures, the integrating period is only a few days. The sensitivity is good enough to enable indoor concentrations to be measured with exposure of 72 h.

1.3 Emanation of radon and thoron from the soil

Uranium and thorium are widely distributed in the earth's crust. Except in geologically recent sediments, there is equilibrium between parent and daughter nuclides in the decay chains leading from uranium and thorium to the radium isotopes (^{224}Ra and ^{226}Ra) which are the precursors of ^{220}Rn and ^{222}Rn. Table 1.1 shows the mean specific activities of the uranium and thorium chains averaged over a worldwide selection of rock samples (Adams *et al.*, 1959). In 327 samples of

Table 1.1. *Abundance of radioactive elements in rocks*

	Uranium chain		Thorium chain	
	(ppm U)	(Bq kg^{-1})	(ppm Th)	(Bq kg^{-1})
Sedimentary rocks	2.5	33	6.5	26
Igneous rocks	3.5	48	13.5	59

surface soil from the USA, Myrick *et al.* (1983) found mean activities of 41 and 35 Bq kg^{-1} for ^{226}Ra and ^{232}Th respectively. There are local variations according to the rock type, and in areas of mineralisation the activities may be up to 1000 times the average.

Typical activities of building materials, such as bricks and concrete, in the UK and USA are in a similar range (20 to 50 Bq kg^{-1}) (Nero, 1983), but granite used in older houses in Cornwall has about 100 Bq kg^{-1} of radium (O'Riordan *et al.*, 1982). In Sweden, Hultqvist (1956) found about 1000 Bq kg^{-1} radium equivalent gamma activity in lightweight concrete made from alum shale.

To measure exhalation of radon a vessel called an accumulator is placed over the soil and sealed to the surface (Wilkening *et al.*, 1972; Keller *et al.*, 1982). To equalise pressures, the vessel is connected to atmosphere or to a bladder by a small vent. The build-up of radon in the accumulator is measured periodically or, in a flow-through arrangement, measurements are made by continuous withdrawal of air to a radon monitor (Schery *et al.*, 1984).

Numerous measurements of ^{222}Rn exhalation have been reported. Global continental means of 15, 16 and 25 mBq m^{-2} s^{-1} have been calculated by Israel (1951), Wilkening *et al.* (1972) and Turekian *et al.* (1977), respectively. Higher rates of emission are found in regions of recent tectonic activity, and where the uranium content of the topmost rocks or soil is enhanced.

The concentration of radium per kg in seawater is only about 10^{-4} of that in rocks. Exhalation of radon from the ocean has been studied to elucidate gas exchange at the surface (Broecker & Peng, 1974). The emission is typically about 0.04 mBq m^{-2} s^{-1}, very small compared with land.

The migration of ^{222}Rn and ^{220}Rn in the soil, and the factors affecting release to atmosphere have been reviewed by Tanner (1964). The energy imparted by the alpha decay of their precursors causes radon atoms to recoil along a track of about 0.03 μm in minerals and about 80 μm in air. Despite the very short range, it seems that release depends more on recoil than on diffusion within the mineral crystal. Emanation to the interstitial air may be primarily from ^{226}Ra or ^{224}Ra on the surface of the crystals.

The emanating power or coefficient of rocks or soil is defined as the proportion of the radon activity per unit bulk volume which is in the interstitial gas. The emanating coefficients of rocks and soils vary greatly. Barretto *et al.* (1972) found values ranging from 0.01 to 0.26 for rocks and from 0.1 to 0.55 for soils. Up to a certain limit, moisture in soil may increase the emanating power. Recoiling radon atoms collide with water molecules in the pore space between grains (Standen *et al.*, 1984), but remain in the interstitial gas as the solubility is low.

Radon moves upwards in soil partly by molecular diffusion in soil gas and partly by bulk flow caused by changes in air pressure at the surface. The diffusion flux can be calculated, if it is assumed that the radio-activity, porosity and density of the soil are independent of depth and that lateral movement of radon can be neglected.

If ρ_s (kg m^{-3}) is the bulk soil density, A_R (Bq kg^{-1}) the specific activity of radium in the soil, α the emanating coefficient, and λ (s^{-1}) the decay constant of radon, then $\rho_s A_R \alpha$ atoms, or $\rho_s A_R \alpha \lambda$ Bq of radon, enter the interstitial air per m^3 of soil volume per second. At depth in the soil, the rates of entry and radioactive decay of radon are equal, so its activity in interstitial air is

$$\chi_\infty = \rho_s A_R \alpha \quad \text{Bq m}^{-3} \text{ of soil} \tag{1.1}$$

Near the surface, radon diffuses upwards. If $\chi(z)$ is the concentration at depth z and D_e the effective diffusion coefficient of radon in soil:

$$\rho_s A_R \alpha \lambda - \lambda \chi(z) + D_e \frac{d^2 \chi(z)}{dz^2} = 0 \tag{1.2}$$

Note that χ is the concentration of radon in the pore spaces per m^3 of soil. The concentration per m^3 of interstitial air in the soil is χ/ε, where ε is the porosity of the soil. Equation (1.2) gives

$$\chi(z) = \chi_\infty [1 - \exp(-z/L)] \tag{1.3}$$

where

$$L = (D_e/\lambda)^{\frac{1}{2}} \tag{1.4}$$

L is termed the diffusion length, and is a measure of the depth of soil from which diffusion of radon is effective. The flux of radon at the surface is

$$Q = D_e \left(\frac{d\chi(z)}{dz} \right)_{z=0} \tag{1.5}$$

$$= \chi_\infty D_e L^{-1} \tag{1.6}$$

$$= \chi_\infty \lambda L \tag{1.7}$$

by substituting D_e from (1.4).

Kraner *et al.* (1964) at Yucca Flat, Nevada, and Schery *et al.* (1984) at Socorro, New Mexico, drew interstitial air from soil at various depths through sampling tubes and measured ^{222}Rn in it. Figure 1.2 shows their results converted to $\chi(z)$ by multiplying by the appropriate values of ε (0.25 at Yucca Flat and 0.35 at Socorro). The curves A and B in Fig. 1.2 are fitted using (1.3) with parameters given in Table 1.2. As it happens, the same value $L = 1.3$ m gives a good fit to both sets of data. Since λ for ^{222}Rn is 2.1×10^{-6} s^{-1}, the corresponding value of D_e from (1.4) is 3.5×10^{-6} m^2 s^{-1}. The exhalation fluxes, deduced from the curves in Fig. 1.2 using (1.7) are 15 and 48 mBq m^{-2} s^{-1} at Yucca Flat and Socorro respectively.

Kraner *et al.* and Schery *et al.* also measured Q directly, by placing accumulator chambers over the ground. Changes in barometric pressure were found to affect Q and also χ at shallow depths. On days of light winds, Kraner *et al.* found that the flux averaged 19 mBq m^{-2} s^{-1}, in fair agreement with the value deduced from the gradient in soil, but on windy days the flux averaged 26 mBq m^{-2} s^{-1}. Over periods of stable barometric pressure, Schery *et al.* at Socorro measured an average Q of

Fig. 1.2. Variation with depth in soil of interstitial radon. ▲, Kraner *et al.*, 1964; △, Schery *et al.*, 1984. Curves A, B are equation (1.3) with $L = 1.3$ m.

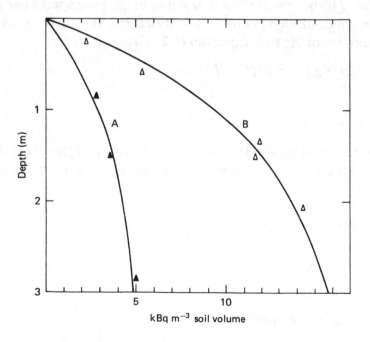

kBq m^{-3} soil volume

Table 1.2. *Exhalation of radon from soil*

Location	Yucca Flat	Socorro
Soil type	Weathered tuff	Gravelly sandy loam
Porosity	0.25	0.35
χ_∞, kBq per m^3 soil	5.4	17.5
L, m	1.3	1.3
Q, mBq m^{-2} s^{-1}		
calculated	15	48
measured	19	34

34 mBq m^{-2} s^{-1}. In the short term, Q varied by a factor 2 due to pressure effects, but Schery *et al.* found that the time averaged exhalation was broadly consistent with the diffusion model.

Both Yucca Flat and Socorro are in arid districts and have porous soils. Lower values of D_e are found in clayey soils (Tanner, 1964, 1980).

As an order of magnitude calculation for average conditions, A_R can be taken as 30 Bq kg^{-1} (Table 1.1). With $\rho_s = 1.5 \times 10^3$ kg m^{-3} and $\alpha = 0.25$, equation (1.1) gives $\chi_\infty = 11$ kBq m^{-3}, intermediate between the values for Yucca Flat and Socorro in Table 1.2. If L is taken as 1 m, equation (1.7) gives $Q = 23$ mBq m^{-2} s^{-1}, which is near the reported global average.

Radon entering buildings comes partly from the underlying soil and partly from building materials. The emanation coefficients and diffusion length in brick and concrete are generally somewhat less than in soil (Ingersoll, 1983), and it is thought that radon in most domestic buildings comes from the soil except where building materials have exceptionally high radium content. Pressure effects are more important in buildings than in the open air because there is usually a small persistent negative pressure in buildings relative to the open air (Nero & Nazaroff, 1984). Pressure variations are a dominant factor in the entry of ^{222}Rn into mine workings, and reductions in concentration have been achieved by maintaining positive air pressure during working, and negative pressure during non-working hours.

The decay constant of ^{220}Rn is 6000 times greater than that of ^{222}Rn, so from (1.4) L is about 80 times less, that is only about 1 cm. The fraction of ^{220}Rn atoms which escape from the rock crystals to the interstitial air is apparently about the same as for ^{222}Rn, and since the specific activities of the thorium and radium chains are similar, equation (1.6) implies that the emanation of ^{222}Rn should be 80 times greater measured in activity, but 80 times less measured in atoms $m^{-2}\,s^{-1}$, than the emanation of ^{220}Rn. Zarcone *et al.* (1986), in a mineralised area of New Mexico, measured exhalation rates of 3.6 and 0.067 Bq $m^{-2}\,s^{-1}$ for ^{220}Rn and ^{222}Rn, a ratio of 54 : 1.

1.4   Radon in air – variations in space and time

Very many measurements of radon in air have been made. Table 1.3 compares results of old measurements (Satterly, 1908, radon absorbed in charcoal and transferred to ionisation chamber) and recent measurement (Keller & Folkerts, 1984, ^{218}Po collected from a chamber followed by alpha spectroscopy). At 15 locations in the UK, Wrixon *et al.* (1988), using passive dosimeters out of doors, found a mean of 3.4 Bq m^{-3}, almost the same as Satterly's result obtained at Cambridge 80 years previously. Year round measurements from four locations in the USA gave average ^{222}Rn in the range 8–12 Bq m^{-3} (Gesell, 1983). Somewhat lower mean levels would be expected in the UK than in USA or Germany, owing to the influence of oceanic air. Most stations show higher mean levels in winter than summer, because vertical dispersion is better in summer. For the same reason, levels are generally higher at night. Over the oceans, and over snow in the Arctic and Antarctic, the radon concentration is about two orders of magnitude lower than over land (Israel, 1951; Lockhart, 1960).

The concentration of radon decreases with height, the gradient being determined by the vertical diffusivity of the atmosphere. Jacobi & André (1963) calculated the gradient by solving numerically the equation

$$\frac{d}{dz}\left(K(z)\frac{d\chi}{dz}\right) - \lambda\chi = 0 \qquad (1.8)$$

where $K(z)$ is the eddy diffusivity at height z. They assumed five characteristic profiles of $K(z)$, corresponding to five categories of

Table 1.3. *Radon in air near ground level*

Location	No. of measurements	^{222}Rn (Bq m^{-3}) Mean	Highest	Lowest	Reference
Cambridge, UK	58	3.9	13	1.3	Satterly (1908)
Saarland, FRG	101	6.7	17	1.1	Keller & Folkerts (1984)

atmospheric stability. They also assumed that the exhalation rate from land is everywhere the same, namely 1 atom $cm^{-2} s^{-1}$ (21 mBq $m^{-2} s^{-1}$) and that horizontal advection of air from regions of lower radon concentration, such as the oceans, could be neglected.

Pearson & Jones (1966) measured radon at various heights near Chicago every 90 min for 8 d, and also measured Q, which, as it happened, was 21 mBq $m^{-2} s^{-1}$, the value assumed by Jacobi & André, Figure 1.3 shows the theoretical gradient in two stability categories (NNN and WNN in Jacobi & André's paper) typical of day and night respectively and the results of Pearson & Jones. Comparison of theory and experiment can only be illustrative, because $K(z)$ depends on surface roughness and wind speed as well as on atmospheric stability. Also χ depends on the average Q over an area upwind, not just on the local value.

Over continental areas, χ continues to diminish with height, and is about 0.1 Bq m^{-3}, at the tropopause (Moore *et al.*, 1973). In the

Fig. 1.3. Radon concentration versus height above ground. $\triangle$, $\blacktriangle$, Pearson & Jones' experimental values during day and during night; A, B, Jacob & André's calculated values for normal turbulence and weak vertical mixing.

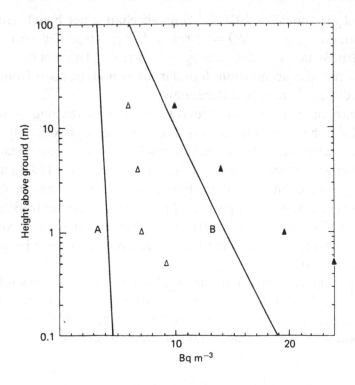

stratosphere, Machta & Lucas (1962) found only 0.02 Bq m^{-3}, the rate of upwards diffusion of radon through the tropopause being slow compared with the radioactive half-life.

1.5   Radon indoors

The concentration of radon is higher indoors than outdoors for three reasons. Air movement is less. Radon is exhaled from walls as well as from the floor. Slight negative pressures indoors, caused by heating and by effects of wind, draw in air from the soil below (Nero & Nazaroff, 1984). There may be continual circulation, whereby air is drawn down through the soil round the building, then up through the floor.

A grossly oversimplified calculation shows the first effect. If χ_1, χ_0 are the concentrations of radon in a groundfloor room and outside, Q the rate of exhalation from the floor, assumed to be the same as from the ground outside, H the height of the room and λ_V the ventilation rate, then neglecting exhalation from the walls

$$Q/H = \lambda_V(\chi_1 - \chi_0)$$
$$(\chi_1 - \chi_0)/Q = (\lambda_V H)^{-1} \tag{1.9}$$

Typically, λ_V is about 3×10^{-4} s^{-1} (one air change per hour), and H about 2.5 m, giving $(\chi_1 - \chi_0)/Q = 1300$ s m^{-1}. If χ_0 is typically 5 Bq m^{-3}, and Q is 20 mBq m^{-2} s^{-1}, this gives $\chi_1 = 31$ Bq m^{-3}. This is sufficient to account for radon concentrations found in many houses, apart from the additional effect of pressure differentials.

A summary of the many recent surveys of radon in dwellings is given in Table 1.4. The distributions were found to be approximately log-normal. In a survey of 300 houses in Cornwall, an area of mainly granitic rocks, Brown et al. (1986) found a median concentration of 170 Bq m^{-3}. The use of granite for walls contributed to the high values in some houses. In earlier work, Hultqvist (1956) found mean concentrations of 15, 47 and 133 Bq m^{-3} respectively in Swedish houses built of wood, brick and concrete. The concrete was made from alum shale having a high radium content.

The Department of the Environment (1988) has published a guide to the reduction of radon levels in homes. Ground floors can be sealed with plastic sheets, or, as a more drastic measure, a sub-floor suction system, exhausting the upwards flow of radon to atmosphere, can be installed.

Table 1.4 Concentration of radon in dwellings

Country	No. of dwellings	^{222}Rn (Bq m^{-3}) Median	Mean	90% percentile	Reference
UK	2309	15	22	43	O'Riordan et al. (1987)
FRG	6000	40	50	90	Schmier & Wicke (1985)
Netherlands	930	24	31	55	Put & de Meijer (1985)
Sweden*	500	69	122	270	Swedjemark & Mjönes (1984)
USA	817	33	55	140	Nero et al. (1986)

* Houses built before 1975.

1.6   Radon in mines

The minimum grade of uranium ore which can be mined profitably by underground workings is about 2000 ppm, three orders of magnitude greater than the average level in crustal rocks. The rates of emanation from the walls of mine workings are correspondingly higher. Evans, (1967) quoted typical Q values of 15 and 18 Bq m^{-2} s^{-1} for American and Soviet mines respectively.

The first surveys of radon in US (Colorado) uranium mines were made in 1952, when the raised incidence of lung cancer in miners first became apparent (United States Public Health Service, 1957). The level was highly variable, with medians for each mine varying from 7×10^2 to 3×10^5 Bq m^{-3}. The median of the medians was 4×10^4 Bq m^{-3}, 1000 times higher than the median concentration in houses (Table 1.4). Subsequently, ventilation was improved and other measures were taken, for example disused parts of mines were sealed off. The median of medians was reduced to 1×10^4 Bq m^{-3} in 1956 and to 4×10^3 Bq m^{-3} in 1966. In a survey of six mines in New Mexico in 1970, George & Hinchliffe (1972) found a median radon concentration of 7×10^3 Bq m^{-3}.

In 1955, the International Commission on Radiological Protection set a maximum permissible occupational concentration of 3.7×10^3 Bq m^{-3} (10^{-10} Ci l^{-1}), for continuous exposure, equivalent to 1.1×10^4 Bq m^{-3} (3×10^{-10} pCi l^{-1}) for a 40-h working week. Subsequently, when it was realised that the critical dose to the lung was from inhalation of decay products, not radon itself, the permissible concentration was defined in terms of the concentration of decay products. The current recommended limit (ICRP, 1986) for a working period of 2000 h per year is 1.5×10^3 Bq m^{-3} equilibrium equivalent radon concentration (a term defined in Section 1.8 below).

1.7   Thoron (^{220}Rn) in air

The radioactive half-life of ^{220}Rn is so short that special measurement techniques have to be used. Israel (1965) found ^{220}Rn concentrations strongly dependent on height, with about 10 Bq m^{-3} at 1 m above ground. Because of the short half-life, the effective source for ^{220}Rn in air is the topsoil over a radius of a few hundred metres from the point of measurement. Comparing ^{220}Rn with ^{222}Rn, the source area is much smaller for ^{220}Rn but the exhalation, in Bq m^{-2} s^{-1}, much larger, and the concentrations in air near the ground are similar in order of magnitude.

As an example of recent measurements, Zarcone *et al.* (1986) used the two filter method (Section 1.2, (c)) with a 36-l chamber. The counts from ^{220}Rn daughters on the end filter were distinguished by alpha spectroscopy. The results (Table 1.5) show the ^{220}Rn/^{222}Rn ratio was 4 outdoors but only 0.15 indoors. The indoor measurements were made in a test house with restricted ventilation, giving opportunity for build-up of ^{222}Rn which did not apply to ^{220}Rn owing to its short half-life. Measurements at a variety of indoor locations by Schery (1985) gave ^{220}Rn/^{222}Rn ratios in the range 0.1 to 1.

High concentrations of ^{220}Rn, in the range 1 to 100 Bq m^{-3}, have been found in factories handling large amounts of thorium (Duggan, 1973; Kotrappa *et al.*, 1976).

1.8 Decay products of radon

The short-lived decay products of ^{222}Rn, namely ^{218}Po (RaA), ^{214}Pb (RaB) and ^{214}Bi (RaC) have radioactive half-lives of 3.05, 26.8 and 19.7 min, respectively. The corresponding decay coefficients are $\lambda_1 = 3.79 \times 10^{-3}$, $\lambda_2 = 4.31 \times 10^{-4}$, $\lambda_3 = 5.86 \times 10^{-4}$ s^{-1}. At height in the atmosphere, these decay products are in equilibrium with ^{222}Rn, but near the ground this is not so. Some of the radon has been exhaled too recently for equilibrium to have been reached. Also, decay products, being isotopes of elements which are solids at normal temperatures, deposit on the ground. They also attach to particles, the socalled condensation nuclei, in the air; this process greatly modifies their diffusivity and mobility.

The dose to the human lung depends less on ^{222}Rn than on the decay products, and the dose to the bronchial epithelium depends particularly on the decay products present as free atoms, molecules, ions or ion clusters as distinct from those attached to condensation nuclei, which are less readily deposited in the respiratory tract. A great deal of work has been done recently on the activities of decay product relative to

Table 1.5. *Radon and thoron in air (Bq m^{-3})*

	^{222}Rn	^{220}Rn	^{220}Rn/^{222}Rn
Outdoors	5.5	22	4.0
Indoors	140	21	0.15

(Zarcone *et al.*, 1986)

radon, and the unattached fraction, in the open air, in dwellings and in mines.

The notation used in the literature for the concentration of decay products in air is confusing. Not only has the unit of activity been changed from curie to becquerel, but three different units are used for the potential alpha energy concentration (*PAEC*). This is the alpha energy delivered if all the radon daughters decay to ^{210}Pb (RaD) and is a measure of the potential dose if the solid products are deposited in the respiratory tract. The dose from the beta and gamma radiation is small in comparison, and becomes negligible when the quality factor (relative biological efficiency) of alpha irradiation is taken into account.

The S.I. unit for the *PAEC* is J m^{-3}, and the unit for the time integrated *PAEC*, or dosage, is Jh m^{-3}. The equivalent radon concentration is that concentration of radon, with decay products in equilibrium, which has the same *PAEC* as have the decay products actually present. In indoor air, the equivalent concentration is often about half the actual radon concentration. High in the atmosphere, the equivalent and actual radon concentrations are the same.

In the uranium mining industry, the Working Level (*WL*) is defined as a concentration of decay products having *PAEC* equal to that of decay products in equilibrium with 100 pCi l^{-1} (3.7×10^3 Bq m^{-3}) of radon. The Working Level Month (*WLM*) is defined as exposure to decay products equivalent to 1 *WL* for 170 h, this being the nominal number of hours worked per month in a mine. The *WLM* is still used in discussions of the epidemiology of lung cancer in relation to exposure to radon and its decay products.

The following notation will be used

χ_0	Activity of ^{222}Rn in air (Bq m^{-3})
χ_1, χ_2, χ_3	Activity of ^{218}Po, ^{214}Pb, ^{214}Bi (Bq m^{-3})
χ_{1u}, χ_{1a}	Activity of ^{218}Po unattached, attached to nuclei (Bq m^{-3})
Q_{1u}, Q_{1a}	Flux of unattached, attached ^{218}Po to surface (Bq m^{-2} s^{-1})
v_u, v_a	Velocity of deposition of unattached, attached ^{218}Po (m s^{-1})
$\lambda_1, \lambda_2, \lambda_3$	Radioactive decay constants of ^{218}Po, ^{214}Pb, ^{214}Bi (s^{-1})
λ_A, λ_V	Rate constants for attachment to nuclei, ventilation (s^{-1})
$\lambda_{Du}, \lambda_{Da}$	Rate constants for deposition of unattached, attached decay products (s^{-1})
E_p	Potential alpha energy concentration (J m^{-3}) of the decay products
E_{pu}	Contribution to E_p from unattached decay products

E_0 Theoretical alpha energy concentration if all decay products were in equilibrium with existing ^{222}Rn (J m^{-3})

F Equilibrium energy ratio $= E_p/E_0$

χ_{eq} Equivalent ^{222}Rn concentration $= F\chi_0$ (Bq m^{-3})

f_u Fraction of ^{218}Po activity or atoms in air not attached to nuclei

f' Ratio of activity of unattached ^{218}Po to activity of ^{222}Rn $= f_u\chi_1/\chi_0$

f_p Fraction of potential alpha energy which is unattached $= E_{pu}/E_p$

f_d Fraction of unattached ^{218}Po or ^{212}Pb deposited in tube

q Rate of formation of small ions (ionised gas molecules) in air (m^{-3} s^{-1})

n Concentration of small ions in air (m^{-3})

N Concentration of condensation nuclei in air (m^{-3})

d_p Diameter of condensation nucleus (m)

α Rate constant for recombination of small ions (m^3 s^{-1})

β Rate constant for attachment to nuclei (m^3 s^{-1}) (Note: $\lambda_A = \beta N$)

D Molecular or Brownian diffusivity (m^2 s^{-1})

k Mobility (m^2 V^{-1} s^{-1})

D_B Dose to bronchial tissue (Gy)

D_{Bu}, D_{Ba} Dose from unattached, attached decay products

H_E Dose-equivalent (Sv).

Conversion factors between S.I. units and those formerly used are given in the Appendix 1.1 at the end of this chapter.

The potential alpha energy per atom of ^{218}Po is 6.0 MeV from its own decay and 7.68 MeV from the decay of ^{214}Po, giving 13.68 MeV or 2.19 pJ. The energies of ^{214}Pb and ^{214}Bi are each 1.23 pJ. The half-life of ^{214}Po is so short that its alpha can be considered as belonging to ^{214}Bi (and the same applies to ^{212}Po in the ^{220}Rn chain). The activity per atom is λ Bq, so the potential alpha energy per Bq is found by multiplying the energy per atom by λ^{-1}, the radioactive mean life in seconds, giving 0.58, 2.86 and 2.10 nJ for ^{218}Po, ^{214}Pb and ^{214}Bi, respectively. The sum of these numbers, 5.54 nJ per Bq, is the ratio E_0/χ_0, and so, for the ^{222}Rn chain:

$$E_p = 0.58\chi_1 + 2.86\chi_2 + 2.10\chi_3 \qquad \text{nJ m}^{-3} \qquad (1.10)$$

$$F = (0.105\chi_1 + 0.52\chi_2 + 0.38\chi_3)\chi_0^{-1} \qquad (1.11)$$

18   Radon

For the ^{220}Rn chain:

$$E_p = 69.1\chi_2 + 6.6\chi_3 \tag{1.12}$$

$$F = 0.913\chi_2 + 0.087\chi_3 \tag{1.13}$$

where χ_2, χ_3 are now the activities (Bq m^{-3}) of ^{212}Pb and ^{212}Bi. The contribution of ^{216}Po (ThA) to E_p is negligible, because its half-life is so short that there are very few atoms per m^3.

To illustrate the values of these quantities, Table 1.6 shows results of Wilkening (1987) obtained over 5 d in an unventilated room at Socorro, New Mexico. The average concentration of ^{222}Rn was 23 Bq m^{-3}, similar to the average level in UK dwellings (Table 1.4). The equilibrium ratio, F, was 0.44 and the potential alpha energy 56 nJ m^{-3}. Assuming 80% occupancy (7000 h a^{-1}), the annual exposure to a person living in the room would be 0.4 mJh m^{-3}, equivalent to 0.11 *WLM*.

Table 1.7 shows examples from various types of location of the relative activity of ^{222}Rn and its decay products, and the equilibrium ratio. Indoors, F depends on the ventilation and on the number of condensation nuclei in the air, since attachment reduces deposition on surfaces, as discussed below.

Table 1.6. *Radon decay products and atmospheric small ions indoors (Wilkening, 1987)*

	Activity (Bq m^{-3})	No. of ions or particles (m^{-3})
Aerosol particles		$1.0 \pm 0.7 \times 10^{10}$
^{222}Rn (χ_0)	23 ± 7	
^{218}Po (χ_1)	22 ± 15	
^{214}Pb (χ_2)	10 ± 4	
^{214}Bi (χ_3)	7 ± 4	
Unattached decay products (small positive ions)		
^{218}Po	1.5 ± 1.3	400 ± 340
^{214}Pb	0.15 ± 0.02	350 ± 50
^{214}Bi	0.03 ± 0.07	50 ± 120
Total		800

$E_p = 0.58 \times 22 + 2.86 \times 10 + 2.10 \times 7 = 56$ nJ m^{-3}
$F = 56/(5.54 \times 23) = 0.44$
$\chi_{eq} = 23 \times 0.44 = 10.1$
$f_u = 1.5/22 = 0.068$
$f' = 1.5/23 = 0.065$
$f_p = (0.58 \times 1.5 + 2.86 \times 0.15 + 2.10 \times 0.03)/56 = 0.024$

Values stated are (means $\pm$ S.D.)

Table 1.7. *Equilibrium between radon and decay products*

Location	Nucleus count (m^{-3})	Relative activity ^{222}Rn	^{218}Po	^{214}Pb	^{214}Bi	Equil. factor F	Reference
Open air	n.s.	100	40	39	39	0.40	Keller & Folkerts (1984)
Dwellings	n.s.	100	45	29	27	0.30	Keller & Folkerts (1984)
Test house	7×10^9	100	41	16	10	0.15	Zarcone et al. (1986)
Uranium mines	7×10^{10}	100	58	25	17	0.25	George & Hinchliffe (1972)

n.s. = not stated.

In the ^{220}Rn chain, ^{216}Po is always in equilibrium, but ^{212}Pb is never so. Israel (1965) found the activity ratio ^{220}Rn/^{212}Pb averaging 120 at 0.5 m height above ground and 80 at 1 m. Indoors, Zarcone *et al.* (1986) found a ratio of 60. In a thorium factory Kotrappa *et al.* (1976) found a median ^{220}Rn/^{212}Pb ratio of 200.

1.9   Radon and thoron decay products as small ions

Although alpha decay carries away positive charge, electrons are stripped from the parent atom by its recoil, and decay products are formed as positive ions. Before discussing their properties, it is convenient to summarise briefly the formation, neutralisation and attachment to condensation nuclei of ordinary small ions in air, as described for example by Chalmers (1967).

Air is ionised by radiation from natural activity in the air and on the ground, and by cosmic rays. Production of one ion pair requires 32.5 eV if ionisation is by fast electrons, 35.6 eV if by alpha rays. The total energy dissipated in air per decay of ^{222}Rn depends on the equilibrium ratio of the decay products. Taking the mean of the results in Table 1.7, it is 10.3 MeV, and the rate of production of ion pairs is approximately

$$q = 3 \times 10^5 \chi_0 \, \text{m}^{-3} \, \text{s}^{-1} \qquad (1.14)$$

For a typical $\chi_0 = 10 \, \text{Bq m}^{-3}$, this gives $q = 3 \times 10^6 \, \text{m}^{-3} \, \text{s}^{-1}$. The total rate of formation of ion pairs from all background radiation, including cosmic rays and gamma rays from soil is normally about $1 \times 10^7 \, \text{m}^{-3} \, \text{s}^{-1}$. It is greater where background is high and in uranium mines may be 100 to 1000 times greater.

In the free atmosphere, the rate of production of small ions is in balance with the rate of neutralisation by recombination and the rate of attachment to condensation nuclei. Condensation nuclei, otherwise called Aitken nuclei, are submicrometric particles mainly produced by combustion processes. If slight differences in the rates of attachment of positive and negative ions are ignored,

$$q = \alpha n^2 + \beta n N \qquad (1.15)$$

where n is the number of small ions of one sign, N the number of condensation nuclei, and α, β are rate constants with units $\text{m}^3 \, \text{s}^{-1}$. Typical values of α and β, derived from measurements of atmospheric electricity, are 1.6×10^{-12} and $1.4 \times 10^{-12} \, \text{m}^3 \, \text{s}^{-1}$ respectively (Junge, 1963).

Figure 1.4 shows the relation between n and N for three values of q. The dashed line separates regions in which αn is greater or less than βN. To the left of the line most ions are neutralised before they are attached. In country air, typically $q = 10^7$ m^{-3} s^{-1}, $N = 1.2 \times 10^{10}$ m^{-3}, $n = 5.5 \times 10^8$ m^{-3}, and the mean life of a small ion before neutralisation $(\alpha n)^{-1}$ is 1100 s, compared with a mean life before attachment $(\beta N)^{-1}$ of 60 s, so most small ions become attached to nuclei, which converts them to large ions. In the clean atmosphere above about 2000 m, N may be only about 10^8 m^{-3}, and most small ions persist as such until neutralised by recombination. In uranium mines, and in most laboratory experiments with ^{222}Rn or ^{220}Rn, q is 10^9 m^{-3} s^{-1} or more, and small ions are rapidly neutralised.

In filtered air, N is zero, $n = (q/\alpha)^{\frac{1}{2}}$ and the mean life before neutralisation is $(\alpha q)^{\frac{1}{2}}$. Using (1.14), and taking $\alpha = 1.6 \times 10^{-12}$ m^3 s^{-1}, the mean life before neutralisation by ions formed by ^{222}Rn and daughters, is about 1400 $\chi_0^{-\frac{1}{2}}$ s. This provides a test of whether the decay products

Fig. 1.4. Relation between number of small ions and nuclei. To left of dashed line, neutralisation is likely to precede capture.

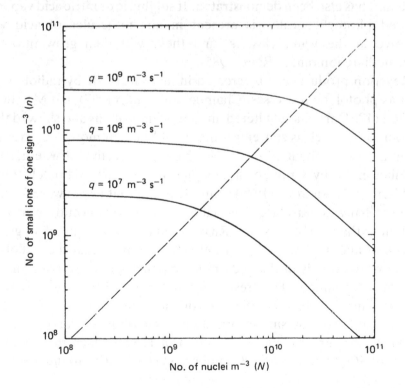

in an experimental arrangement are likely to have been ions or neutral atoms.

In air containing water vapour, mass spectrometry indicates that positive ions are mostly hydrated protons, $H^+(H_2O)_n$, where n may be any number between 1 and about 8 (Shahin, 1966; Huertas *et al.*, 1971). Negative ions are probably mostly hydrated O^- or OH^-. The formation of clusters of water molecules round ions is very rapid, but in unpolluted air the clusters do not grow beyond about 1 nm diameter and remain as small ions until they become attached to condensation nuclei. They then become large ions.

Large ions themselves can be classified in two size ranges. In the terminology of Whitby (1978), the nuclei mode centred on 10 nm is distinct from the accumulation mode, centred on 100 nm. In urban air, the number of particles in the nuclei mode is greater than the number in the accumulation mode, but their total surface area is less, and it is surface area which determines the probability of attachment of small ions to particles with diameters of order 100 nm or less. Thus the large ions are mostly in the accumulation mode. The distinction between small and large ions is well established in atmospheric electricity. Nevertheless, the existence of intermediate ions, in the size range 1–10 nm, has also been demonstrated. If sulphuric or nitric acid vapour, formed photochemically, is present in air, molecules of acid will dissolve in the water clusters, and these will then grow into the intermediate ion range (Raes, 1985).

Reaction products can be created in laboratory air by radiolytic as well as photolytic processes. Chamberlain *et al.* (1957) and Megaw & Wiffen (1961) irradiated filtered air in flowing systems and showed that condensation nucleus concentrations of 10^{10} m^{-3} or more were created when a dose of about 10 mGy was given, irrespective of whether the irradiation was by X-rays, beta- or alpha-rays. Bricard & Pradel (1968), and Kruger & Andrews (1976) similarly found that nuclei were formed when ^{220}Rn was introduced into laboratory air at concentrations such that more than 10^{10} atoms of ^{220}Rn decayed to ^{212}Pb per m^3 of air giving a dose of about 20 mGy. Bricard & Pradel showed that the radiolytic nuclei were initially uncharged, indicating that ions were not required for their formation. The freshly formed nuclei had a diffusivity of 7×10^{-7} m^2 s^{-1}, corresponding to a diameter of 3 nm.

A possible mechanism for formation of radiolytic nuclei is radiolysis of water vapour, leading to formation of H_2O_2, which then oxidises traces of SO_2 to give H_2SO_4. Chamberlain *et al.* (1979) found increased formation of radiolytic nuclei in laboratory air when SO_2 was added. Addition of O_3 to the air also increased nucleus production, whereas addition of NO, a well-known radical scavenger, inhibited it.

1.10   Mobility of decay products

The mobility of an ion in air is its velocity under an electric field. The mobility and Brownian diffusivity of ions determine their rate of attachment to condensation nuclei and to surfaces.

Jonassen & Hayes (1972) measured the mobility of radon daughter products in an unventilated basement laboratory where the ^{222}Rn concentration was about 250 Bq m^{-3}. The central electrode of a cylindrical capacitator was covered with aluminium foil, and maintained at negative voltage. When air was drawn through the cylinder, decay products were deposited at different distances along the foil according to their mobility. The results (Fig. 1.5) showed mobility varying from 0.35 to 2.5×10^{-4} m^2 V^{-1} s^{-1} with the mode at 0.5 and the mean at 0.94×10^{-4} m^2 V^{-1} s^{-1}. Similar measurements in the open air at Socorro, New Mexico had shown a similar range with mean at

Fig. 1.5. Mobility distribution of ^{222}Rn daughters. I, in open air (Wilkening *et al.*, 1966); II, in laboratory (Jonassen & Hayes, 1972).

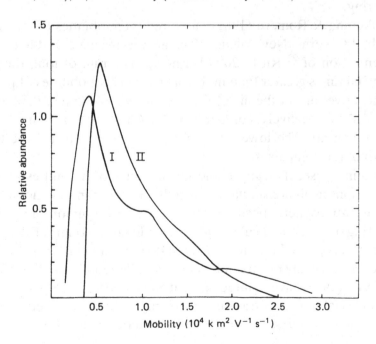

1×10^{-4} m^2 V^{-1} s^{-1} (Wilkening *et al.*, 1966). Bricard *et al.* (1965), in laboratory experiments with air artificially enriched with decay products, found mobilities in the range 0.3 to 1.2×10^{-4} m^2 V^{-1} s^{-1}, with a small separate distribution, comprising a few per cent of the total, at about 2×10^{-4} m^2 V^{-1} s^{-1}. Fontan *et al.* (1969), using filtered air, found mobilities in four discrete groups, ranging from 0.55 to 2×10^{-4} m^2 V^{-1} s^{-1}, the proportion in the highest group decreasing with the time elapsed since the ion was formed.

In experiments in argon at low pressure, Munson *et al.* (1939) found that the mobility of alkali ions was diminished by clustering when water vapour (2.8%) was added. However, the effect diminished with increasing atomic weight of the alkali ions, and the mobility of Cs$^+$ was reduced only from 2.23 to 2.18×10^{-4} m^2 V^{-1} s^{-1} by adding water vapour to argon. In pure water vapour, however, the mobility of Cs$^+$ was only 0.7×10^{-4} m^2 V^{-1} s^{-1}. Munson *et al.* attributed the effect of atomic weight to diminished ability to attract water molecules as the size of the ion increased, and to a diminished effect on mobility per molecule attracted. The lower mobilities in pure water vapour, which applied to all alkali ions, were explained by the greater frequency of collisions with water molecules. The ageing time of the ions in these experiments was of order milliseconds or less, and, at the low vapour pressures employed, the number of collisions may have been a limiting factor in clustering.

Wilkening & Romero (1981) measured the mobilities of ions in the Carlsbad Caverns, New Mexico. The air was stagnant and had a high concentration of ^{222}Rn ($\sim$2000 Bq m^{-3}). The rate of ionisation was about 200 times greater than in the open air. The mobilities of positive and negative ions in the air of the caverns were 0.35 and 0.50×10^{-4} m^2 V^{-1} s^{-1} respectively, compared with 1.6 and 2.0×10^{-4} m^2 V^{-1} s^{-1} in the open air. The lowered mobility was thought to be due to the growth of ion complexes.

To summarise, the ranges and mean values of mobilities of decay product ions in air are similar to, or rather lower than, the mobilities of ordinary atmospheric positive ions. A possible common factor is the clustering of water molecules, but this would not account for the ageing effect observed by Fontan *et al.* (1969) since the clustering of water molecules normally occurs in less than a millisecond. Raes (1985) and Raes *et al.* (1985) have pointed out that air in laboratory apparatus often has appreciable concentrations of organic vapours, derived from the materials used, unless care is taken to exclude them. Trace molecules may be attached to the small ions or may take part in radiolytic reactions giving products which are attached.

1.11 Diffusivity of decay products

The diffusivity D $(\text{m}^2 \text{ s}^{-1})$ of a molecule or molecular cluster is related to the mobility k $(\text{m}^2 \text{ V}^{-1} \text{ s}^{-1})$ of the singly charged ion at NTP by

$$k = 39.6 \ D \tag{1.16}$$

It is important to know the diffusivities of decay products, whether as ions, neutral atoms or molecules, or as cluster ions, in order to calculate deposition in the respiratory tract. One way of doing this is to measure deposition on the walls of a tube in laminar flow.

The fraction deposited is related to the diffusivity by Gormley & Kennedy's (1949) equation

$$\begin{aligned} f_d = 1 &- 0.819 \exp(-7.3\,\xi) - 0.0975 \exp(-44.6\,\xi) \\ &- 0.035 \exp(-114\,\xi) \end{aligned} \tag{1.17}$$

where x (m) is the length of the tube, q $(\text{m}^3 \text{ s}^{-1})$ is the flow down it, and $\xi = \pi D x / 2q$

Figure 1.6 shows the arrangement used by Chamberlain & Dyson (1956). Filtered air with ^{222}Rn or ^{220}Rn was passed into a chamber, where the residence time was several minutes, and decay products were formed, and thence down the tubes T_1 and T_2, which were lined with filter paper and closed at the end by the filters F_1 and F_2. The activities found on T_2 and F_2 were subtracted from those on T_1 and F_1 to correct the latter for the contributions of decay products formed within T_1. This correction was only a few percent when ^{222}Rn was used, and was negligible when ^{220}Rn was used, because nearly all the ^{220}Rn decayed in the chamber. This illustrates the fact that decay products inhaled as such, and not those formed in situ, contribute most of the dose to the airways when radon is inhaled.

In these experiments, and those of Raabe (1968), the levels of airborne activity, and hence of ionisation in the chamber, were sufficient to ensure that decay products were neutralised, by collision with negative air ions, before entering the diffusion tube.

Figure 1.7 shows f_d from equation (1.17) for three values of D, and Chamberlain & Dyson's experimental points. In further experiments, the relative humidity of the air was varied from 18 to 88%, but no effect on f_d for neutral ^{212}Pb was observed (Chamberlain, 1966). In experiments with diffusion tubes, the fit to equation (1.17) is not perfect, and Busigin *et al*. (1981) showed that it often depends on the range of values of the parameter ξ (equation 1.17) in the particular experiment. The value $D = 5.4 \times 10^{-6}$ m^2 s^{-1}, which has been widely quoted, was chosen by Chamberlain & Dyson to correspond to a mobility $k = 2.17 \times 10^{-4}$ m^2 V^{-1} s^{-1}, deduced by reference to Munson *et al.*'s

Fig. 1.6. Measurement of diffusivity of decay products.

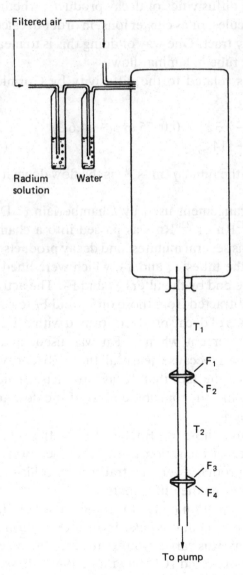

(1939) measurements of the mobility of unclustered thallium ions in nitrogen.

A different type of diffusion sampler, in which decay products are deposited as they pass from an orifice between concentric plates, was developed by Mercer & Stowe (1969), and was used by Raghunath & Kotrappa (1979) to measure D for decay products of both ^{222}Rn and ^{220}Rn, with results shown in Table 1.8. The activities of ^{220}Rn and ^{222}Rn were in the range 0.4 to 3×10^6 Bq m^{-3}, high enough to ensure neutralisation, but also high enough to create radiolytic products when the ageing time in the chamber was extended by reducing the airflow. In humidified but otherwise pure argon no nuclei were created and the

Fig. 1.7. Deposition of unattached ^{212}Pb in tubes. Flow rates (cm^3 s^{-1}), 83.3, 83.3, 33.3, 16.6, 8.3; corresponding tube diam. (cm), 1.8, 0.4, 1.8, 0.4, 1.8; symbols, ○, △, ×, ●, +. Curves A, B, C are equation (1.17) with $D = 6.5$, 5.4 and 4.5×10^{-6} m^2 s^{-1}, respectively.

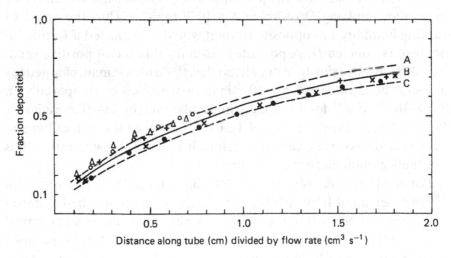

Table 1.8. *Diffusivity (m^2 s^{-1} $\times$ 10^6) of ^{218}Po and ^{212}Pb in air*

Ageing time in chamber (min)	^{218}Po		^{212}Pb	
	RH = 10%	RH = 90%	RH = 10%	RH = 90%
1	8.1	8.0	5.5	5.4
10	6.5	4.4	0.8*	0.7*

*Condensation nuclei present.
Data from (Raghunath & Kotrappa, 1979).
RH = relative humidity.

ageing effect was much less. The ageing effect was also observed by Fontan *et al*. (1969), who found D for neutral ^{212}Pb declining from 9 to 2×10^{-6} m^2 s^{-1} as the ageing time in filtered laboratory air increased from 0.5 to 20 min.

In uranium mines, ionisation of the air is probably sufficient to neutralise the radon decay products, but in outside air most are present as ions. Thomas & Le Clare (1970) developed a radon monitor of the two-filter type, an arrangement similar to that of Fig. 1.6, and calculated the fraction of ^{218}Po generated in the lower tube (T$_2$) which deposited on its walls. They found that experiment agreed with theory if D for ^{218}Po was 8.5×10^{-6} m^2 s^{-1}. The concentration of ^{222}Rn ranged from 2×10^4 to 3×10^5 Bq m^{-3} and the stay time in the tube from 3 s to 1 min. At the lower end of these ranges, the ^{218}Po was presumably present as ions.

Variation of the relative humidity of the air showed little effect until it was less than 10%, when the proportion of ^{218}Po reaching the end filter increased, implying D of about 5×10^{-6} m^2 s^{-1}. Thus the effect of varying humidity was opposite to what would be expected if formation of clusters affected D. A possible reason for this is that positive small ions diffuse more slowly in dry air, so that ^{218}Po ions remain charged for longer. The D value of 8.5×10^{-6} in normal air may be compared with 7.6×10^{-6} m^2 s^{-1} for I$_2$ vapour, molecular weight 254 (Krzesniak & Porstendörfer, 1978). Thomas & Le Clare passed air through a filter and carbon bed (gas mask canister) before it entered the apparatus, thus probably eliminating organic vapours.

Porstendörfer & Mercer (1979) did similar experiments with ^{220}Rn-laden air at 10^6 to 10^9 Bq m^{-3}. Their diffusion tube had a central electrode and deposition was measured with and without an electrical field. Collection on the charged electrode was more efficient in moist than in very dry air (relative humidity, $RH < 2\%$). In moist air, D was 6.8×10^{-6} m^2 s^{-1} irrespective of whether the ^{212}Pb was partially or wholly neutralised before deposition. In very dry air and low ^{220}Rn concentrations, D was 4.7×10^{-6}, and it was concluded that the charged component had D equal to 2.4×10^{-6} m^2 s^{-1}. Paradoxically, this would correspond to $k = 1 \times 10^{-4}$ m^2 V^{-1} s^{-1}, the mobility found by Jonassen & Hayes (1972) for ^{222}Rn decay products in moist air. In all Porstendörfer & Mercer's experiments, the ageing time was very short.

Busigin *et al*. (1981) considered that values of D for ^{218}Po reported in the literature depend on the ratio of flow rate to length of the diffusion tubes, or on the age of the ion. They measured the efficiency of electrostatic collection in various pure gases and concluded that ^{218}Po, or its oxide, was neutralised by charge transfer from gases of low ionisation potential such as nitric oxide. Frey *et al.* (1981) measured D of ^{218}Po in various gases by Thomas & Le Clare's method and obtained values ranging from 7.9×10^{-6} in moist N_2 to 3.1×10^{-6} in dry N_2 with 10 ppm NO. They concluded that the lower values applied to ions and the higher to atoms neutralised by charge transfer. At what concentration of pollutant gas, charge transfer becomes important is not known, but if Wilkening's results in Table 1.6 are typical, it does not have the effect of reducing substantially the lifetime of ^{218}Po ions in normal air.

Careful reviews by Raes (1985) and Raes *et al.* (1985) leave unanswered the question of the role of humidity, and of acid or organic vapours, in modifying the diffusivity of decay product ions. By comparison with the mobility in normal air of decay product and ordinary atmospheric small ions, the diffusivity of decay product small ions is probably 2 to 3×10^{-6} m^2 s^{-1}. For neutral atoms, or possibly oxide molecules, most measurements give D in the range 5 to 8×10^{-6} m^2 s^{-1}, except where radiolytic reaction products or reactive trace gases are present in sufficient concentration to form intermediate ions.

1.12 Attachment of decay products to nuclei

Attachment is mainly by diffusion, although, if the decay products are ions, electrostatic attraction to charged nuclei of opposite sign makes a small additional effect (Bricard & Pradel, 1966). The rate constant for attachment λ_A, is given by an equation originally applied to evaporation from small droplets (Fuchs, 1959).

$$\lambda_A = \frac{4\pi r N D}{D(rV_m\alpha)^{-1} + r(r+\Delta)^{-1}} \tag{1.18}$$

where N is the number of condensation nuclei per unit volume of air, r is their radius, Δ is the mean free path of decay product molecules, α is the accommodation coefficient or sticking probability of decay products on nuclei and V_m is the component of mean kinetic velocity of vapour molecules perpendicular to a surface. $V_m = (RT/2\pi M)^{\frac{1}{2}}$, where R is the gas constant, T absolute temperature and M molecular weight of decay products.

Δ is about 15 nm, and if r is much less than this the second term in the denominator of (1.18) is small and

$$\lambda_A = 4\,\pi r^2 N V_m \alpha \qquad\qquad (1.19)$$

which is the rate of collision between molecules and particles multiplied by the sticking probability. λ_A is then proportional to the surface area of the condensation nuclei.

For particles of about 1 μm radius, $r(r + \Delta)^{-1}$ is near unity. Either term in the denominator of (1.18) may then be dominant, depending on whether $D(rV_m\alpha)^{-1}$ is greater or less than unity. At room temperature and pressure, V_m for a decay product atom is 43 m s^{-1} and if $D = 7 \times 10^{-6}$ m^2 s^{-1} and $r = 1$ μm, the condition $D(rV_m\alpha)^{-1} > 1$ is equivalent to $\alpha < 0.14$. Thus if α is less than 0.1, its value determines λ_A, which then depends on r^2 up to $r = 1$ μm.

As the particle size increases further, α becomes unimportant unless it is very small and (1.18) reduces to Smoluchowski's equation

$$\lambda_A = 4\pi r N D \qquad\qquad (1.20)$$

Fig. 1.8. Attachment coefficient of decay product ions. $\times$, Kruger & Andrews, 1976; $+$, Kruger & Nothling, 1979; $\bigcirc$, Porstendörfer & Mercer (1978); $\triangle$, Porstendörfer *et al.* (1979). Line is equation (1.18) with $D = 7 \times 10^{-6}$ m^2 s^{-1}.

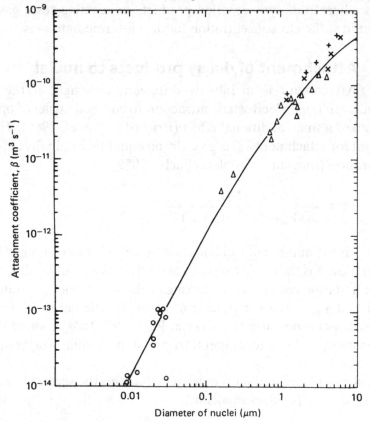

The rate of attachment is then determined by the rate of diffusion through the boundary layer round the particle.

The attachment coefficient β of equation (1.15) is λ_A/N. Figure 1.8 shows β from (1.18) with $D = 7 \times 10^{-6}$ m^2 s^{-1}, $V_m = 44$ m s^{-1} and α unity, and also shows experimental points of Kruger & Andrews (1976), Kruger & Nothling (1979), Porstendörfer & Mercer (1978) and Porstendörfer et al. (1979), obtained by measuring the rate of attachment to monodisperse aerosols. Porstendörfer & Mercer found no appreciable difference in the attachment of ^{212}Pb whether as atoms or ions. Kruger & Nothling found β to be slightly greater for ^{222}Rn than for ^{220}Rn decay products.

Figure 1.9 shows Junge's (1963) natural aerosol size distribution, typical of well-populated country districts, as the cumulative percentage by number (curve A) and by volume (curve B) less than given diameter. The number median is 0.06 μm (Junge's distribution includes particles such as sea salt and resuspended dust which extend the distribution at the large-diameter end). By numerical integration, using Fig. 1.8,

Fig. 1.9. Size distributions of condensation nuclei. A, Junge's number distribution; B, Junge's mass distribution; C, activity distribution of newly attached decay products.

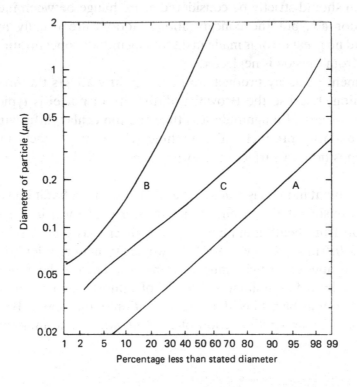

$\lambda_A = 2.1 \times 10^{-2}$ s^{-1}, and since N for the Junge aerosol is 1.7×10^{10} m^{-3}, the corresponding value of β is 1.2×10^{-12} m^3 s^{-1}. Measured values of β, for room and outdoor aerosols, average 1.4×10^{-12} m^3 s^{-1} (Porstendörfer, 1984). This is also the textbook value for attachment of ordinary atmospheric small ions to nuclei.

The calculated size distribution of newly attached decay products is shown as curve C in Fig. 1.9. The activity median diameter is 0.16 μm. With passage of time, the distribution would be shifted to larger particle sizes, as coagulation proceeds. George (1972) used diffusion batteries to measure the size distribution of nuclei carrying radon decay products and found activity median diameters (AMD) averaging 0.18, 0.11, and 0.30 μm in a city basement, fifth floor room, and rural outside air, respectively.

1.13   Deposition of decay products on surfaces

Decay products are formed in air by the decay of their precursors. They are removed by three processes:

(a) their own radioactive decay,
(b) ventilation,
(c) deposition on surfaces.

Ventilation should strictly be considered as exchange between indoor and outdoor air, but the concentrations indoors are usually much higher, and no great error is made in discussing indoor concentrations if the input from outside is neglected.

Attachment of decay products to nuclei greatly affects the process of deposition, because the Brownian diffusivity of nuclei is typically about four orders of magnitude less than the molecular diffusivity of unattached decay products. The lifetime of decay products in air before deposition on surfaces is shorter if the air is clean than if it is dirty.

Air movement indoors is much slower than outdoors, but it is usually enough to ensure that concentrations are fairly uniform in a room. Convection from heating appliances gives air speeds typically in the range 0.05–0.5 m s^{-1} (Daws, 1967). However, to undergo deposition, vapour molecules or particles must be transported across the boundary layer, typically a few millimetres thick, of almost stagnant air over surfaces. This may be achieved by sedimentation, molecular or Brownian diffusion, or under the action of electrostatic or thermophoretic forces.

The velocities of deposition of attached and unattached decay products are the fluxes divided by the concentrations

$$v_a = Q_a/\chi_a, \qquad v_u = Q_u/\chi_u \qquad (1.21)$$

If the velocities of deposition are averaged over all the surfaces, and S (m^{-1}) is the ratio of surface area to volume in a room, the rate constants for removal by deposition are

$$\lambda_{Da} = v_a S, \qquad \lambda_{Du} = v_u S \qquad (1.22)$$

λ_{Da} and λ_{Du} will depend on the air movement in the room, the area and nature of the surfaces, and the strength of any electrostatic fields. For λ_{Da}, the size distribution of the nuclei in the air, and the presence or absence of thermal gradients near surfaces may also be important. Despite the variables, some experimental and observational data are available which allow λ_{Da} and λ_{Du} to be compared in order of magnitude with the radioactive decay constants and with the rate constant for ventilation.

In the simple case of airflow over an aerodynamically smooth surface, with a fully developed boundary layer, the velocity of deposition can be calculated as a function of the diffusivity of the vapour or particle and the air speed. Formulae, developed for mass and heat transfer (Brutsaert, 1982) have been shown to apply to both attached and unattached ^{212}Pb in wind tunnel experiments (Chamberlain, 1966, 1968; Chamberlain *et al.*, 1984).

Figure 1.10 shows the velocities of deposition to a smooth surface of attached and unattached decay products, with a range of possible values of D. The scales are μm s^{-1} for the attached and mm s^{-1} for the unattached decay products, illustrating the effect of attachment on diffusivity. For a nucleus of unit density with diameter, d_p, equal to $0.17\,\mu$m, the sedimentation velocity is 2 μm s^{-1}, and deposition by Brownian diffusion and by sedimentation to upwards-facing surfaces are of comparable efficiency. For smaller particles, Brownian diffusion is always more effective.

Surfaces are not all smooth, and the fetch of air movement over them may be insufficient for full development of the boundary layer. Wind tunnel experiments with moderately rough surfaces, such as textiles (Chamberlain, 1966), indicate that v_a might be increased by a factor of two or three relative to the value for smooth surfaces.

Knutson *et al.* (1983) analysed measurements of the plate-out of attached decay products in an experimental chamber, and obtained a range of values of v_a with median 7.5 μm s^{-1}. Jacobi (1972) devised a model of the formation, attachment to nuclei, and plate-out of decay products, and applied it to the conditions in uranium mines. Reineking *et al.* (1985) and Porstendörfer *et al.* (1987) similarly used Jacobi's model to analyse measurements of ^{222}Rn and its decay products in houses, and found them consistent with λ_{Da} in the range 3×10^{-5} to 1×10^{-4} s^{-1} (0.1 to 0.4 h^{-1}). Assuming a surface/volume ratio about 3 m^{-1}, the corresponding values of v_a are 10 to 30 μm s^{-1}, somewhat higher than expected from Fig. 1.10 or from Knutson *et al.*'s (1983) measurements. Thermophoretic forces may contribute significantly to deposition in living rooms. Deposition of airborne dust is often observed to cause stains on walls near central heating appliances. Electrostatic effects may also contribute, since most building materials and furnishings are dielectrics and can carry electrostatic charge.

Fig. 1.10. Velocity of deposition of attached (left-hand scale) and unattached (right-hand scale) decay products to smooth surfaces. Curves A, B, C, D, E; corresponding d_p (μm), 0.17, 0.12, 0.08, molecular, molecular; D (m^2 s^{-1}), 3×10^{-10}, 5×10^{-10}, 1×10^{-9}, 0.05, 0.07.

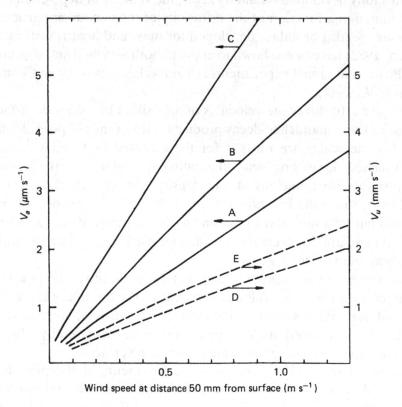

Wind speed at distance 50 mm from surface (m s^{-1})

Ventilation in centrally heated dwellings, though much less than in those with open fires, usually gives 0.1 to 0.3 air changes per hour, so ventilation and deposition on surfaces are about equally effective in limiting the concentration of attached decay products in dwellings.

Unattached decay products plate-out much more rapidly. From Fig. 1.10, v_u can be expected to be of the order 1 mm s^{-1} in rooms with normal air movement. Bigu (1985) measured plate-out of unattached decay products of both ^{222}Rn and ^{220}Rn in a chamber of volume 26 m^3. He found v_u varying from 0.9 to 2.4 mm s^{-1} (^{222}Rn products) and from 0.6 to 5.3 mm s^{-1} (^{220}Rn products) depending on whether or not a fan was operated in the chamber.

Vanmarcke *et al.* (1987) found that a value $v_u = 1.9$ mm s^{-1} gave the best fit in model calculations applied to their measurements in dwellings. If $v_u = 2$ mm s^{-1} is taken as a reference value, with a surface/volume ratio of 3 m^{-1}, then $\lambda_{Du} = 6 \times 10^{-3}$ s^{-1} = 20 h^{-1}.

Table 1.9 shows typical values of the rate coefficients for unattached ^{218}Po in dwellings. The nucleus concentration is taken as 1.7×10^{10} m^{-3} with a Junge-type distribution of d_p, as described above (section 1.12).

1.14 Potential alpha energy of unattached decay products

When radon is inhaled, the dose to the bronchial passages from deposited decay products is much greater than the dose from the radon itself, and depends on the potential alpha energy of the readily deposited unattached decay products. Per Bq, the potential alpha energy of ^{214}Pb is greater than that of ^{218}Po, but, per atom, that of ^{218}Po is greater. Recoil may detach ^{214}Pb from a nucleus at the moment of its formation (Mercer, 1976), but the rate of formation of ^{218}Po, atoms per second, is usually greater than that of ^{214}Pb (Table 1.7). If the mean lives of the two nuclides as free atoms are similar, there are more free

Table 1.9. *Typical rate coefficients for removal of unattached ^{218}Po*

	Rate coefficient	
	s^{-1}	h^{-1}
Radioactive decay, λ_1	3.7×10^{-3}	13
Ventilation, λ_V	3×10^{-4}	1
Attachment to nuclei, λ_A	2×10^{-2}	70
Plate-out, λ_{Du}	6×10^{-3}	20

^{218}Po than ^{214}Pb atoms. Hence most of the unattached potential alpha energy, E_{pu}, is associated with ^{218}Po.

By considering the rate at which ^{218}Po atoms become attached to nuclei, and the rates of removal from air, it is readily shown that the fraction of ^{218}Po activity which is unattached is

$$f_u = \frac{\lambda_1 + \lambda_V + \lambda_{Da}}{\lambda_1 + \lambda_A + \lambda_V + \lambda_{Da}} \tag{1.23}$$

Note that λ_{Du} does not enter this expression. The ratio of the activity of unattached ^{218}Po to the activity of ^{222}Rn is

$$f' = \frac{\lambda_1}{\lambda_1 + \lambda_A + \lambda_V + \lambda_{Du}} \tag{1.24}$$

and the mean life of ^{218}Po in the unattached state is

$$f'\lambda_1^{-1} = \frac{1}{(\lambda_1 + \lambda_A + \lambda_V + \lambda_{Du})} \tag{1.25}$$

In dwellings, and in mines except where the ventilation rate is very high, λ_1 is much greater than λ_V or λ_{Da} (though not necessarily greater than λ_{Du}), and (1.23) becomes

$$f_u = \frac{\lambda_1}{\lambda_A + \lambda_1} = \frac{\lambda_1}{\beta N + \lambda_1} \tag{1.26}$$

Figure 1.11 shows experimental measurements of f_u obtained both indoors and outdoors. Agreement with equation (1.26) is fair, but measurements of f_u with $N > 10^{11}$ m^{-3} tend to exceed the theoretical value, probably because the nuclei were derived from local sources and the median diameters were less than the value assumed in estimating β.

The unattached part E_{pu} of the total potential alpha energy E_p is $f_p E_p$ by definition of f_p. It can also be related to the concentration of ^{222}Rn since

$$E_{pu} = Ff_p E_0 = 5.54\, Ff_p \chi_0 \qquad \text{nJ per Bq} \tag{1.27}$$

The product Ff_p gives the unattached $PAEC$ per Bq of radon, and is important in assessing the dose to the bronchial epithelium, which depends largely on the unattached activity.

Figure 1.12 shows model values of f_p and F as used by James

Fig. 1.11. Fraction of ^{218}Po unattached to nuclei. $\triangle$, Duggan & Howell, 1969; $\times$, $+$, George, 1972, indoors and outdoors, respectively; $\blacktriangle$, Reineking et al., 1985; $\bigcirc$, George & Hinchliffe, 1972. Line is equation (1.26) with $\beta = 1.4 \times 10^{-12}$ m^3 s^{-1}, $\lambda_1 = 3.8 \times 10^{-3}$ s^{-1}.

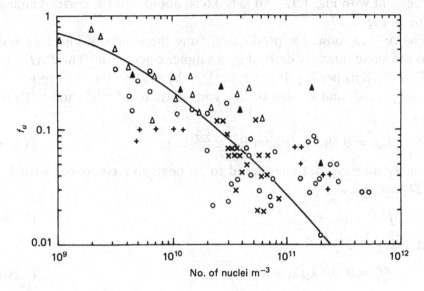

Fig. 1.12. Equilibrium ratio, F, and fraction of $PAEC$ unattached, f_p (James, 1987a, after Porstendörfer, 1984).

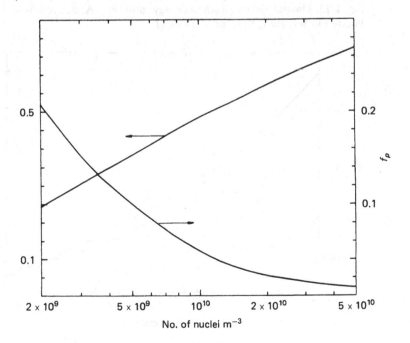

(1987a,b), derived from Porstendörfer's (1984) analysis. As N increases, F increases but f_p decreases.

In the example of Table 1.6, F is 0.44 with $N = 1 \times 10^{10}$ m^{-3}, in fair agreement with Fig. 1.12, but f_p is 0.024, about half the corresponding value in Fig. 1.12.

Figure 1.13 shows the product Ff_p from the results of Fig. 1.12 and also the same quantity derived by a simpler calculation. The $PAEC$ of ^{218}Po is 0.58 nJ per Bq. If losses of ^{218}Po by ventilation and deposition are neglected, and so also are the contributions of ^{214}Pb and ^{214}Bi to E_{pu},

$$E_{pu} = 0.58 f' \chi_0 \text{ nJ per Bq of } ^{222}\text{Rn} \tag{1.28}$$

Since λ_V and λ_{Du} are considered to be negligible compared with λ_1, (1.24) becomes

$$f' = \lambda_1 (\lambda_1 + \lambda_A)^{-1} \tag{1.29}$$

and, using (1.27)

$$Ff_p = 0.105\, \lambda_1 (\lambda_1 + \lambda_A)^{-1} \tag{1.30}$$

Figure 1.13 shows Ff_p from (1.30), with λ_A equated to $1.4 \times 10^{-12}\ N$.

Fig. 1.13. Unattached equilibrium energy ratio, Ff_p. A, derived from Porstendörfer, 1984; B, from equation (1.30).

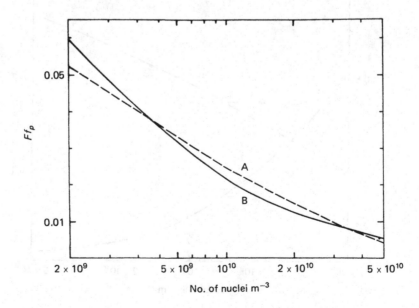

No. of nuclei m^{-3}

There is little difference over most of the range compared with Ff_p derived from Fig. 1.12.

Chamberlain & Dyson (1956) assumed that ^{218}Po (RaA) would have a mean life of 30 s before atachment to nuclei, giving $\lambda_A = 3.3 \times 10^{-2}$ s^{-1}. With $\lambda_1 = 3.79 \times 10^{-3}$ s^{-1}, equation (1.30) then gives $Ff_p = 0.011$, a value corresponding to $N = 2.5 \times 10^{10}$ m^{-3} in Fig. 1.13.

In the thoron (^{220}Rn) chain, E_p is determined by the airborne lifetime of ^{212}Pb, and E_{pu} by its lifetime as a free atom or molecule. The ratio indoors of radon/thoron in the example of Table 1.5 is 0.15, and in measurements at 25 other indoor locations Schery (1985) found the activity ratio to be 0.23. If the ratio is 0.2, this is also the ratio of the number of atoms of ^{212}Pb to the number of atoms of ^{218}Po formed per m^{3}s. The alpha energy per disintegration of ^{212}Pb, allowing for the branching decay of ^{212}Bi, is 7.85 MeV compared to 13.68 MeV for ^{218}Po. Hence, if the mean lives as free atoms are similar, E_{pu} of the thoron chain should be only about 10–15% of the E_{pu} of the radon chain. However, Schery measured the total potential alpha energy, E_p, of both chains at 68 indoor and 16 outdoor locations and found mean values of $E_p(^{220}$Rn$)/E_p(^{222}$Rn$)$ of 0.62 indoors and 0.29 outdoors. These results seem to imply longer airborne lifetimes for ^{212}Pb than for ^{218}Po $+ ^{214}$Pb. The conclusion may be invalidated by the fact that the locations where Schery measured the Rn parent isotopes were fewer and less diverse than the locations where E_p was measured.

1.15 Radiation dose from inhalation of radon and decay products

When radon and its decay products are inhaled, some of the decay products are deposited in the respiratory tract. Most of the active deposit remains in the lung for several hours, during which time the short-lived daughters of ^{222}Rn undergo decay. However the half-life of ^{212}Pb (ThB) is long enough for substantial transfer to the bloodstream to take place before decay (Booker *et al.*, 1969; Hursh *et al.*, 1969), and this reduces the dose from ^{220}Rn.

To calculate the radiation dose it is necessary to know:

(a) the volume of air breathed per hour,
(b) the fraction of inhaled decay products deposited in the respiratory tract,
(c) the distribution between various parts of the tract and the modifications produced by movement in the mucous flow and uptake into blood.

Figure 1.14 shows the fractional retention in the respiratory tract of particles inhaled through the mouth, and also the retention in the tracheobronchial (T–B) region only, according to calculations by Yu & Diu (1982) and Egan & Nixon (1987). Also shown are experimental values found in volunteer experiments by Chamberlain *et al.* (1978), Tu & Knutson (1985), and Schiller *et al.* (1988). The fraction deposited increases from 0.1 to 0.9 as d_p decreases from 0.3 to 0.03 μm. Over this range of size, sedimentation and Brownian diffusion are the mechanisms of deposition, and the fractional deposition depends on the stay time in the lung. The calculated lines in Fig. 1.14 are for a breathing cycle of 15 breaths/min, and a tidal volume of 1 l. In estimating the radiation dose, separate consideration must be given to attached and unattached decay products.

From Fig. 1.14, if decay products are attached to nuclei of diameter 100 nm, about 25% of inhaled activity is deposited in the lung. This would be reduced to about 19% if the tidal volume were halved, but increased to 33% if the duration of the breathing cycle were doubled (Egan & Nixon, 1987). Figure 1.14 refers to mouth breathing, but

Fig. 1.14. Fractional deposition of particles in the respiratory tract, and in the T-B region of the tract (15 breaths/min by mouth, tidal volume, 1 l). Calculations of Yu & Diu, 1982, full lines; and Egan & Nixon, dashed lines. Experimental results of: Tu & Knutson, 1984, ●; Schiller *et al.*, 1988, ○; Chamberlain *et al.*, 1978, ×.

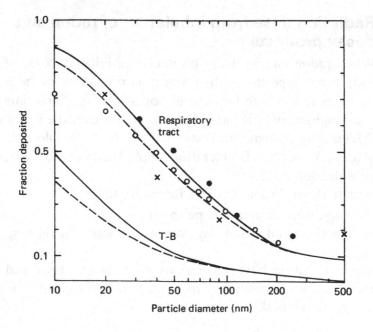

submicrometric particles are not removed efficiently in the nasal passages (George & Breslin, 1969) and to calculate the radiation dose to the
lung in average conditions, 20% deposition in the pulmonary region can
be assumed.

If the rate of ventilation of the lungs is 0.75 m^3 h^{-1}, the rate of
deposition of potential alpha energy in the alveolar region is then 0.15
E_p Jh^{-1}. The mass of the lung in the standard man is 1 kg, and by
definition 1 J kg^{-1} is 1 Gy, so the dose rate is 0.15 Gy per Jh m^{-3}. Curve
A of Fig. 1.15, following James (1987b), shows the variation of the
alveolar dose with d_p. As d_p decreases, the dose increases to a peak at 18
nm, and then declines because smaller particles are trapped in the upper
respiratory tract and fewer reach the alveolar region.

Unattached decay products have diffusivity appropriate to d_p about
1 nm, and are almost completely retained in the respiratory tract

Fig. 1.15. Dose rate to alveolar (A) and bronchial (B) regions of lungs by
inhalation of radon daughters (James, 1987b). Range of values for bronchial
dose reflects uncertainties in breathing pattern, airway size and clearance
rates.

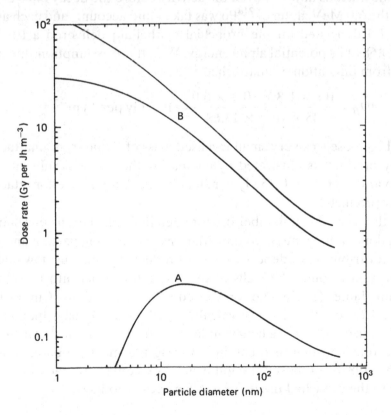

(Booker *et al.*, 1969). When volunteers inhaled decay products via the nose about 50% deposition was found in the nasal passages alone (George & Breslin, 1969), and this reduced the fraction reaching the lung.

The critical tissue is the bronchial epithelium, where radiation-induced lung cancers are believed to originate, and many calculations and experiments with models of the upper respiratory system have been done to estimate the dose to this region (James, 1987,a,b).

Chamberlain & Dyson (1956) measured deposition of unattached ^{212}Pb in a rubber model of the trachea and main bronchi. At an inspiratory flow of 20 l min^{-1}, corresponding to a ventilation rate of about 0.6 m^3 h^{-1}, the activity deposited per cm^2 of surface in the bronchi was 0.18% of that entering the trachea. No difference was found in the deposition of unattached ^{218}Po (RaA) compared with unattached ^{212}Pb (ThB).

Chamberlain & Dyson assumed that the alpha energy was deposited in bronchial tissue to a depth of 45 μm, the range of ^{218}Po alpha particles, and neglected absorption in the mucous layer. They assumed that the mucous flow removed the activity before the decay of ^{214}Pb, so only the 6.0 MeV alpha of ^{218}Po was taken into account, and each atom of ^{218}Po deposited on the bronchial epithelium delivered a fraction 6.0/13.68 of its potential alpha energy. With these assumptions the dose rate from inhalation of unattached ^{218}Po is

$$D_B = \frac{0.6 \times 1.8 \times 10^{-3} \times 6.0}{45 \times 10^{-7} \times 13.68} = 105 \text{ Gy per Jh m}^{-3} \qquad (1.31)$$

The high dose to a very small localised mass of tissue from unattached decay products is illustrated by comparing the above result with the alveolar dose rate of 0.15 Gy per Jh m^{-3} calculated above for attached decay products.

In the past 30 a, a number of more detailed calculations of D_B have been done, taking into account variations in breathing pattern and also the redistribution of deposited decay products by mucous flow and by absorption through the walls of the respiratory tract into the bloodstream. James (1987a,b) has reviewed these calculations. Curves B of Fig. 1.15 show the range of calculated values of D_B as a function of particle size, assuming a lung ventilation rate of 0.75 m^3 h^{-1}. The upper curve refers to mouth breathing. Taking the most probable values, James (1987b) proposed the following dose conversion factors for the dose to the bronchial tissue from ^{222}Rn decay products:

Unattached: $D_{Bu} = 37$ Gy per Jh m^{-3} (1.32)

Attached: $D_{Ba} = 2$ Gy per Jh m^{-3} (1.33)

Since the *WLM* is 3.5 mJh m^{-3} (Appendix 1.1), the factors are 130 mGy (unattached) and 7 mGy (attached) per *WLM*. In the USA, the National Council on Radiological Protection and Measurements (1984) proposed factors of 140 and 6 mGy per *WLM*.

The conversion factor for the bronchial dose from both unattached and attached decay products is

$$D_B = f_p D_{Bu} + (1 - f_p) D_{Ba} \text{ Gy per Jh m}^{-3} \tag{1.34}$$

Taking f_p indoors as 0.05 (Reineking *et al.*, 1985), and D_{Bu}, D_{Ba} from (1.32), (1.33), the contributions to the dose from free and attached decay products are similar and:

$$D_B = 3.75 \text{ Gy per Jh m}^{-3} \tag{1.35}$$

$$= 13 \text{ mGy per } WLM \tag{1.36}$$

In mines, nuclei are numerous and f_p is low, for example a median value of 0.01 can be derived from measurements of George & Hinchliffe (1972). Equation (1.34) with $f_p = 0.01$ gives 2.35 Gy per Jh m^{-3} (8.8 mGy per *WLM*), with most of the dose coming from the attached fraction.

A number of other factors enter into the comparison of dose. The average miner has a higher ventilation rate than a person at home. Against this, the bronchial dose depends on the particle size of the nuclei (Fig. 1.15), and George *et al.* (1975) found this to average 0.17 μm in mines compared with 0.12 μm in dwellings. The US National Research Council Committee (1988) have suggested a conversion factor of 8 mGy per *WLM* for miners, with considerable possible variation depending on ventilation rate, and other factors.

In industrial exposure, continuously operated filter samplers can be used to measure the potential alpha energy concentration, but in houses passive dosimeters are more commonly used and these measure concentrations of radon, not radon daughters. Since

$$E_p = E_0 F = 5.54 \chi_0 F \text{ nJ m}^{-3} \tag{1.37}$$

the conversion factor from bronchial dose in terms of radon is

$$D_B = 5.54 F \left[f_p D_{Bu} + (1 - f_p) D_{Ba} \right] \text{ nGy per Bqh m}^{-3} \tag{1.38}$$

In Figure 1.16, D_B is shown as a function of N, assuming F and f_p from Fig. 1.12 and D_{Ba}, D_{Bu} from (1.32) and (1.33). For a typical indoor nucleus concentration, averaged over day and night, of 1×10^{10} m^{-3}, the conversion factor is 10 nGy per Bqh m^{-3}. Possible variations in the number and size of nuclei, the ventilation of the dwelling, and the breathing pattern of the occupants, imply an uncertainty of at least a factor two either way, in this estimate. Chamberlain & Dyson's (1956) calculation, which only took into account D_{Bu}, was equivalent to a conversion factor of 6.2 nGy per Bqh m^{-3} for a mouth-breathing subject.

The dose to the alveolar region of the lung depends only slightly on the unattached component. If $F = 0.5$, the *PAEC* is 2.72 nJ m^{-3}, per Bq m^{-3} of ^{222}Rn. The conversion factor 0.15 Gy per Jh m^{-3}, calculated above for $d_p = 100$ nm, then gives an alveolar dose of 0.4 nGy per Bqh m^{-3} of ^{222}Rn. The dose to fatty tissue, from dissolved ^{222}Rn, is not more than 5 pGy per Bqh m^{-3}. The significance of relatively high dose rates in very small volumes of tissues is a matter of debate. If it is accepted that the bronchial epithelium is the critical tissue, an important parameter in evaluating indoor exposure is the mean life of ^{218}Po before attachment to nuclei, as suggested by Chamberlain & Dyson

Fig. 1.16. Dose to bronchial tissue from decay products in terms of concentration of ^{222}Rn.

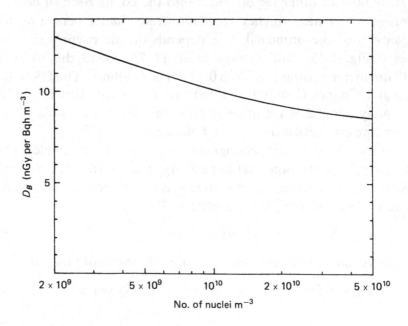

(1956). Since the number of nuclei is usually less in dwellings than in uranium mines, the bronchial dose, per Bq m^{-3} of ^{222}Rn, is greater in the domestic than in the industrial situation.

The potential alpha energy of decay products of ^{220}Rn (i.e. principally ^{212}Pb) in indoor air is approximately half that of decay products of ^{222}Rn (Schery, 1985). Moreover, the radioactive half-life of ^{212}Pb is long enough for considerable absorption into the bloodstream, and movement in mucous flow to occur before the alpha-emitting ^{212}Bi is formed, and the dose to bronchial epithelium per E_p inhaled, is only about a third of that from ^{222}Rn decay products (James, 1987b).

1.16 Dose equivalent

The biological effectiveness of dose depends on the type of radiation and also on the mass and sensitivity of the irradiated tissue. For alpha irradiation, a quality factor of 20 is assumed (ICRP, 1981), and the dose in Sieverts is 20 times the dose in Grays. In addition, ICRP recommends a weighting factor of 0.12 for irradiation of the whole lung and 0.06 for irradiation of bronchial epithelium only. Thus the 'effective dose equivalent', symbol H_E, is defined as the dose to the whole body which carries the same risk as the given dose to the organ or tissue. This, for irradiation of bronchial tissue is $20 \times 0.06 = 1.2$ times the dose to the organ in Gy.

With this convention, the effective dose equivalent in the average indoor atmosphere is given in terms of the *PAEC* by

$$H_E = 3.75 \times 1.2 = 4.5 \text{ Sv per Jh m}^{-3} \tag{1.39}$$

$$= 16 \text{ mSv per } WLM \tag{1.40}$$

UNSCEAR (1982) adopted a lower figure, 5 mSv per WLM.

The ICRP Task Group (1988) has recommended a conversion factor of 1.0×10^{-5} mSv per Bqh m^{-3} of equilibrium equivalent decay product concentration, with a possible 30% uncertainty either way depending on the value of f_p. Since 1 Bq m^{-3} equilibrium equivalent gives 5.54 nJ m^{-3} (Appendix 1.1) the ICRP recommendation is equivalent to 1.8 Sv per Jh m^{-3} or 6.3 mSv per *WLM*.

Different assumptions on the volume of the bronchial tissue considered as critical in calculating the dose, and on the fraction of the *PAEC* which is carried by unattached daughter products, account for the variation in the estimates.

In terms of the radon concentration, 10 nGy per Bqh m^{-3}, derived above from equation (1.38), are equivalent to 12 nSv per Bqh m^{-3}. In calculating the annual dose, it is assumed that a person spends 7000 h per annum indoors, so the annual effective dose equivalent is about 80 μSv per Bq m^{-3}.

The National Radiological Protection Board (1990) have adopted a slightly lower figure, 50 μSv per Bq m^{-3} (coincidentally equivalent to Chamberlain & Dyson: 6.2 nGy per Bqh m^{-3}). For the average UK indoor concentration of 20 Bq m^{-3} (Table 1.4), this gives 1 mSv annually, which is about equal to the total dose from all other natural sources of radiation (Clarke & Southwood, 1989).

1.17 Risk of cancer from inhalation of radon and decay products

The International Commission on Radiological Protection (ICRP, 1981) estimated the likelihood of incurring cancer from irradiation (excluding genetic effects) as 1.25×10^{-2} per Sv of effective dose equivalent (a figure likely to be revised upwards as a result of recent studies). Subsequently (ICRP, 1988) they considered the risk from lifetime indoors exposure to decay products as 2×10^{-8} per Bqh m^{-3} equilibrium equivalent activity in each year. These two estimates are broadly compatible, as follows. From (1.37), the *PAEC* is 5.54 nJ m^{-3} per Bq m^{-3} of equilibrium equivalent activity, and from (1.39) the 50-a effective dose equivalent to the bronchial tissue is $5.54 \times 4.5 \times 50$ nSv = 1.25 μSv. Hence the lifetime risk, per Bqh m^{-3} in each year is $1.25 \times 10^{-6} \times 1.25 \times 10^{-2} = 1.6 \times 10^{-8}$.

Jacobi & Paretzke (1985) estimated that the uranium miners in Colorado accumulated an average exposure of 820 *WLM* in the years 1950–77. Using the factor 16 mSv per *WLM*, the average bronchial dose would have been 13 Sv, giving a 20% chance of cancer on the basis of the ICRP (1981) estimate. The BEIR IV report of the US National Research Council (1988) recorded 256 deaths from lung cancer among the Colorado miners. The total exposure was 73 600 person-years, and 58 deaths would have been expected if there were no carcinogenic effects.

For exposure indoors of the general population, it is more useful to work in terms of the concentration of radon than the concentration of decay products. Brown *et al.* (1986) found the average concentration of ^{222}Rn in Cornish houses to be 300 Bq m^{-3}, and Nero (1988) has estimated that 2% of homes in the USA have similar or higher concentrations. Assuming an annual effective dose equivalent of 80 μSv per Bq m^{-3}, as in the previous section, the lifetime dose is 1.2 Sv. Thus occupants of many homes receive doses which are a tenth of those received by the Colorado miners, and on the above analysis carry a 2% chance of lung cancer. A more recent estimate, taking into account a higher cancer risk factor and also the synergistic effect of smoking, gives a 4.5% risk of lung cancer to UK residents in homes with 300 Bq m^{-3} (National Radiological Protection Board, 1990). As Nero (1988) has pointed out, such a high level of risk greatly exceeds those normally associated with environmental hazards.

1.18 Lead 210 and its daughter products

The short-lived decay products of ^{222}Rn decay to ^{210}Pb, which has a radioactive half-life of 22 a, and decays to ^{210}Bi and ^{210}Po as follows.

$$^{210}\text{Pb} \rightarrow {}^{210}\text{Bi} \rightarrow {}^{210}\text{Po} \rightarrow {}^{206}\text{Pb (stable)}$$

$$22\,\text{a} \qquad 5.0\,\text{d} \qquad 138\,\text{d}$$

A small proportion of the atoms of ^{222}Rn exhaled from the earth's surface are redeposited as short-lived decay products, but most decay in the atmosphere to ^{210}Pb. The mean residence time of ^{210}Pb is short compared with its radioactive half-life, so the world-wide downwards flux of ^{210}Pb, in atoms m^{-2} s^{-1}, should equal the upwards flux of ^{222}Rn. Deposition of ^{210}Pb varies from about 25 Bq m^{-2} a^{-1} (800 atoms m^{-2} s^{-1}) at island sites to 250 Bq m^{-2} a^{-1} (8000 atoms m^{-2} s^{-1}) at continental sites (Turekian *et al.*, 1977). At Milford Haven, Wales, the fallout was found to be 85 Bq m^{-2} a^{-1} (Peirson *et al.*, 1966). Calculations by Turekian *et al.* (1977) show a best fit to the measured fluxes if the average continental exhalation of ^{222}Rn is 1.2×10^4 atoms m^{-2} s^{-1}, (25 mBq m^{-2} s^{-1}).

Table 1.10 shows measurements of ^{210}Pb in air at ground level. The continental effect is apparent, concentrations being lower where the climate is oceanic. It has been suggested that ^{210}Pb was formed in nuclear bomb tests, but there is no evidence that concentrations were higher in the period (1955–65) of frequent bomb tests. Indoor concentrations have not been reported. Blanchard (1969) found ^{210}Pb concentrations in uranium mine air to be very variable, with median about 1 Bq m^{-3} compared with about 10^4 Bq m^{-3} for ^{222}Rn.

^{210}Pb was found to increase with height in the atmosphere of the UK from 0.21 mBq kg^{-1} of air at ground level to 0.26 at 7.6 km and 0.36

mBq kg^{-1} in the stratosphere (Peirson *et al.*, 1966). In the mid-west USA, Moore *et al.* (1973) found a different profile, the concentration decreasing with height to 0.04 mBq kg^{-1} at the tropopause but increasing again to 0.28 mBq kg^{-1} in the stratosphere. The differences in the profiles reflect the geographical situations, the lower part of the profile over the UK being depleted by deposition during passage of air westwards over the Atlantic.

1.19   Residence time in atmosphere

The activities of ^{210}Pb, ^{210}Bi and ^{210}Po can be used to estimate the residence time, T_R, of aerosols in atmospheric air. It is assumed that the three nuclides are distributed similarly with respect to particle size and are affected similarly by removal mechanisms. Table 1.11 shows the integrated vertical profile, that is the total activity of ^{210}Pb above unit area of ground up to the tropopause, and also the deposition fluxes, in the UK and mid-west USA. These results give T_R equal to 8 and 4 d, respectively. Since relatively more of the ^{210}Pb is in the lower atmosphere in the USA, a higher removal rate is to be expected.

Table 1.12 shows the activity ratios ^{210}Bi/^{210}Pb and ^{210}Po/^{210}Pb as measured in air and in rain. If the air in the troposphere is considered as

Table 1.10 *^{210}Pb in air at ground level*

Location	Date	mBq m^{-3}	Reference
Berkshire, UK	1958	0.26	Peirson *et al.*, 1966
Berkshire, UK	1961–5	0.23	Peirson *et al.*, 1966
Freiburg, FRG	1959	0.37	Lehmann & Sittkus, 1959
USA, 6 sites	1966	0.69	Magno *et al.*, 1970
Puerto Rico	1966	0.31	Magno *et al.*, 1970
Hawaii		0.17	Magno *et al.*, 1970
USA, 4 sites	1974–6	0.60	Feely *et al.*, 1981
USA, Wisconsin	1979	0.37	Talbot & Andren, 1983
USA, 9 sites	1985–6	0.75	Graustein & Turekian, 1986
Melbourne, Australia	1966–70	0.16	Bonnyman *et al.*, 1972

a well-mixed reservoir, calculation of the build-up and removal of the daughter nuclides gives

$$^{210}\text{Bi}/^{210}\text{Pb} = T_R/(T_E + T_R) \tag{1.41}$$

$$^{210}\text{Po}/^{210}\text{Pb} = T_R^2/(T_R + T_E)(T_R + T_F) \tag{1.42}$$

where T_E and T_F are the radioactive mean lives of ^{210}Bi and ^{210}Po, namely 7.2 and 199 d respectively.

The Bi/Pb ages of 5 to 8 d found by Fry & Menon (1962) and Moore *et al.* (1972, 1973) are comparable with the results of Table 1.11 and also with the T_R value of 9 d observed for the disappearance of ^{137}Cs from the atmosphere after the Chernobyl accident (Fig. 2.8). In Hawaii, with oceanic air, the Bi/Pb ratio indicates a longer residence time.

The Po/Pb derived residence lives ranging from 19 to 50 d are longer than the Bi/Pb lives. Possible explanations are as follows:

(a) Incursions of stratospheric air
A small proportion of air derived from a long-residence reservoir will affect the apparent Po/Pb age more than the Bi/Pb age. For example, if the mixture is 85% tropospheric with $T_R = 5$ d and 15% stratospheric with $T_R = 200$ d, the apparent Bi/Pb age is 7 d and the apparent Po/Pb age 22 d.

(b) Resuspension of soil
In topsoil, ^{210}Pb and ^{210}Po are in equilibrium, so resuspended soil would increase the apparent age of air samples. However, this is unlikely to be the explanation at island or coastal sites such as Hawaii and Milford Haven. Talbot & Andren (1983) used a cascade impactor to sample air at a mid-continental site (Wisconsin). The $^{210}\text{Po}/^{210}\text{Pb}$ ratio on the impactor stages was about unity, compared with 0.43 on filter collections, suggesting a large particle ($>1\,\mu\text{m}$) component with $^{210}\text{Po}/^{210}\text{Pb}$ in

Table 1.11. *Integrated vertical profile of ^{210}Pb and fallout rate*

Location	Date	^{210}Pb (Bq m^{-2})	(Bq m^{-2} d^{-1})	Mean life (d)	Reference
UK	1958	1.9	0.23	8	Peirson *et al.*, 1966
USA	1970–1	1.5	0.38	4	Moore *et al.*, 1973

Table 1.12. *Isotopic ratios and residence times*

Sample	Bi/Pb		Po/Pb		Reference
	Ratio	T_R (d)	Ratio	T_R (d)	
Air, Freiburg, FRG	0.07	—	0.07	20	Lehmann & Sittkus, 1959
Rain, Freiburg, FRG	—	—	0.17	50	
Air, Milford Haven, Wales	—	—	0.15	40	Peirson et al., 1966
Rain, Arkansas	0.46	6	—	—	Fry & Menon, 1962
Rain, Arkansas	0.74	20	0.15	40	Gavini et al., 1974
Air, Colorado	0.42	5	0.066	20	Moore et al., 1972, 1973
Rain, Colorado	0.53	8	0.062	19	Moore et al., 1972, 1973
Air, Hawaii	0.85	41	0.087	25	Moore et al., 1972, 1973
Air, stratosphere	0.90	745	0.43	170	Moore et al., 1972, 1973

equilibrium. However, the activity median diameters (AMD) of ^{210}Pb and ^{210}Po were 0.4 and 0.65 μm respectively, hardly more than would be expected for an accumulation mode aerosol. For example George (1972) found an AMD of 0.30 μm for the short-lived decay products of ^{222}Rn at Sterling Forest, NY.

Another possibility is the volatilisation of ^{210}Po from soil or vegetation. In the laboratory, polonium is notorious for its ability to transfer from one surface to another. Abe & Abe (1969) found that 60% of ^{210}Po collected on a filter paper volatilised on heating to 200°C for one hour. The inventory of ^{210}Po in topsoil is about 5000 times the inventory in air, when both are expressed per unit area of ground, so only a small degree of volatilisation or resuspension would materially increase the concentration in air.

(c) Volcanic activity

Lambert *et al.* (1982) measured the concentrations of ^{210}Pb and ^{210}Po in the plumes of 11 volcanoes, and found ^{210}Po highly enriched relative to ^{210}Pb, on account of its greater volatility, the median ^{210}Po/^{210}Pb ratio being about 40. Enriched ^{210}Po was found downwind at great distances from volcanoes, the ^{210}Po/^{210}Pb ratio being still above unity 2000 km from Mt Erebus. Lambert *et al.* considered that volcanoes may contribute materially to the global ^{210}Po budget, estimated to be 4×10^{22} atoms or 2.4 PBq per year.

1.20 Release of ^{210}Po in accident to Windscale reactor

Bismuth was irradiated in the Windscale reactor to make ^{210}Po for use in nuclear weapons, and in the reactor fire of October 1957, an estimated 8.8 TBq (240 Ci) of ^{210}Po was released to atmosphere (Crick & Linsley, 1983). The fact that ^{210}Po was released was published at the time (Stewart & Crooks, 1958; Blok, 1958).

The dosage of ^{210}Po in air was measured by analysis of air filters from a number of towns in the UK and continental Europe. The highest measured dosage was 1.6×10^5 Bqs m^{-3} at Liverpool, but calculation indicates that a dosage of 2.6×10^6 Bqs m^{-3} might have been reached in the zone of maximum exposure 6 km downwind of the reactor. This compares with a natural annual ^{210}Po dosage of about 10^4 Bqs m^{-3}.

The adult effective dose equivalent from inhaling the ^{210}Po was estimated at 2 mSv (Crick & Linsley, 1983). This is comparable with the effective dose equivalent from indoor ^{222}Rn and its decay products, but the organs in the body receiving the highest dose from ^{210}Po are the liver and kidneys, not the lung. The average annual effective dose equivalent from natural ^{210}Pb and its daughter products is estimated by UNSCEAR (1982) to be 130 μSv.

Appendix 1.1 Units for radon and daughter products in air

$$1 \, Ci = 3.7 \times 10^{10} \, Bq$$

$$1 \, MeV = 1.6 \times 10^{-13} \, J$$

As explained in Section 1.8, potential alpha energy of daughter products in equilibrium with 1 Bq is 5.54×10^{-9} J (^{222}Rn), 7.57×10^{-8} J (^{220}Rn). So for ^{222}Rn,

$$E_p = 5.54 \times 10^{-9} \, \chi_{eq} = 5.54 \times 10^{-9} \, F\chi_0 \quad J \, m^{-3}$$

$$1 \, WL = E_0 \, (100 \, pCi \, l^{-1} \, of \, ^{222}Rn)$$
$$= E_0 \, (3.7 \times 10^3 \, Bq \, m^{-3} \, of \, ^{222}Rn)$$
$$= 2.05 \times 10^{-5} \, J \, m^{-3}$$

$$1 \, WLM = 170 \times 1.05 \times 10^{-5} \, J \, m^{-3} = 3.5 \times 10^{-3} \, J \, m^{-3}$$

The equivalent radon concentration (χ_{eq}) is that concentration of ^{222}Rn with decay products in equilibrium, which has the same potential alpha energy concentration (PAEC) as have the decay products actually present. Hence, if $\chi_{eq} = 1 \, Bq \, m^{-3}$, $PAEC = 5.54 \, nJ \, m^{-3}$.

Appendix 1.2 Units of radiation dose

Unit	Symbol	Definition
Roentgen	R	X or γ radiation producing 1 ESU of ions of each sign per cm^3 of dry air at STP. Equivalent to 84 erg g^{-1} (8.4×10^{-3} J kg^{-1}) in air.
Rad	rad	Absorbed dose of 100 erg g^{-1} (1×10^{-2} J kg^{-1}) in tissue.
Rem	rem	Dose in rad multiplied by relative biological efficiency (quality factor).
Gray	Gy	Absorbed dose of 1 J kg^{-1}. 1 Gy = 100 rad = 6.24×10^{12} MeV kg^{-1}.

Unit	Symbol	Definition
Sievert	Sv	Dose equivalent, equal to dose in Gy multiplied by quality factor (unity for X, γ and β rays, 10 for neutrons, 20 for α particles)
Effective dose equivalent	Sv or Sv_{eff}	Dose equivalent in Sv multiplied by weighting factor (0.12 for lung, 0.03 for thyroid gland).

Appendix 1.3 Symbols for multiples of 10

Multiple	Symbol	Multiple	Symbol
10^{18}	E	10^{-3}	m
10^{15}	P	10^{-6}	μ
10^{12}	T	10^{-9}	n
10^{9}	G	10^{-12}	p
10^{6}	M	10^{-15}	f
10^{3}	k	10^{-18}	a

References

Abe, S. & Abe, M. (1969) Volatility of ^{210}Po (RaF) in airborne ducts at various temperatures. *Health Physics*, **17**, 340–1.

Adams, J.A.S., Osmond, J.K. & Rogers, J.J.W. (1959) The geochemistry of thorium and uranium. In: *Physics and Chemistry of the Earth* 3, New York, Pergamon Press, pp. 298–348.

Barretto, P.M.C., Clark, R.B. & Adams, J.A.S. (1972) Physical characteristics of radon 222 emanations from rocks, soils, and minerals. *Natural Radiation Environment II*, ed. J.A.S. Adams, W.M. Lowder & T.F. Gesell, Springfield, Va. NTIS pp. 731–40.

Bigu, J. (1985) Radon daughter and thoron daughter deposition velocity, and unattached fraction under laboratory-controlled conditions and in underground uranium mines. *Journal of Aerosol Science*, **16**, 157–65.

Blanchard, R.L. (1969) Radon-222 daughter concentrations in uranium mine atmospheres. *Nature*, **223**, 287–9.

Blok, J. (1958) Airborne radioactivity after a reactor accident. *Proceedings 2nd International Conference on Peaceful Uses of Atomic Energy*, **18**, 325–8. Geneva, United Nations.

Bonnyman, J., Duggleby, J.C. & Molina Ramos, J. (1972) Lead-210 in the Australian environment. In: *Natural Radiation Environment II*, ed. J.A.S. Adams, W.M. Lowder & T.F. Gesell, Springfield, Va. NTIS, pp. 819–931.

Booker, D.V., Chamberlain, A.C., Newton, D. & Stott, A.N.B. (1969) Uptake of radioactive lead following inhalation and injection. *British Journal of Radiology*, **42**, 457–61.

Bricard, J., Girod, P. & Pradel, J. (1965) Spectre de mobilité des petits ions radioactifs de l'air. *C.R. Acad. Sci. Paris*, **260**, 6587–90.

Bricard, J. & Pradel, J. (1966) Electric charge and radioactivity. In: *Aerosol Science*, ed. C.N. Davies, New York, Academia Press, pp. 91–104.

(1968) Formation and evolution of nuclei of condensation that appear in air initially free of aerosols. *Journal of Geophysical Research*, **73**, 4487–96.

Broecker, W.S. & Peng, T.H. (1974) Gas exchange rates between air and sea. *Tellus*, **26**, 21–5.

Brown, L., Green, B.M.R., Miles, J.C.H. & Wrixon, A.D. (1986) Radon exposure of the United Kingdom population. *Environment International*, **2**, 45–8.

Brutsaert, W.H. (1982) *Evaporation into the Atmosphere*. Dordrecht, Reidel.

Busigin, A., van der Vooren, A.W., Babcock, J.C. and Phillips, C.R. (1981) The nature of unattached RaA (^{218}Po) particles. *Health Physics*, **40**, 333–43.

Chalmers, J.A. (1967) *Atmospheric Electricity*, 2nd edn, Oxford, Pergamon.

Chamberlain, A.C. (1966) Transport of gases to and from grass and grass-like surfaces. *Proceedings of the Royal Society A*, **290**, 236–65.

(1968). Transport of gases to and from surfaces. *Quarterly Journal of the Royal Meteorological Society*, **94**, 318–32.

Chamberlain, A.C. & Dyson, E.D. (1956) The dose to the trachea and bronchi from the decay products of radon and thoron. *British Journal of Radiology*, **29**, 319–25.

Chamberlain, A.C., Heard, M.J., Little, P., Newton, D., Wells, A.C. & Wiffen, R.D. (1978) Investigations into lead from motor vehicles. Atomic Energy Research Establishment Report 9198, London, HMSO.

Chamberlain, A.C., Heard, M.J., Penkett, S.A. & Wells, A.C. (1979) Suppression of radiolytic nuclei in air by nitric oxide. *Health Physics*, **37**, 706–7.

Chamberlain, A.C., Garland, J.A. & Wells, A.C. (1984) Transport of gases and particles to surfaces with widely spaced roughness elements. *Boundary-layer Meteorology*, **29**, 343–60.

Chamberlain, A.C., Megaw, W.J. & Wiffen, R.D. (1957) Role of condensation nuclei as carriers of radioactive particles. *Geofiscia Pura et Applicata*, **36**, 233–42.

Clarke, R.H. & Southwood, T.R.E. (1989) Risks from ionising radiation. *Nature*, **338**, 197–8.

Crick, M.J. & Linsley, G.S. (1983) An assessment of the radiological impact of the Windscale reactor fire of October, 1957. National Radiological Protection Board, Chilton. Report R-135, Addendum.

Daws, L.F. (1967) Movement of air streams indoors. In: *Airborne Microbes*, ed. P.H. Gregory & J.L. Monteith, Cambridge University Press, pp. 31–59.

Department of the Environment (1988). *The Householder's Guide to Radon*. London, HMSO.

Duggan, M.J. (1973) Some aspects of the hazard from airborne thoron and its daughter products. *Health Physics*, **24**, 301–10.

Duggan, M.J. & Howell, D.M. (1969) Relationship between the unattached fraction of airborne RaA and the concentration of condensation nuclei. *Nature*, **224**, 1190–1.

Egan, M.J. & Nixon, W. (1985) A model of aerosol deposition in the lung for use in inhalation dose assessments. *Radiation Protection Dosimetry*, **11**, 5–17.

(1987). Mathematical modelling of fine aerosol deposition in the respiratory system. In: *Deposition and Clearance of Aerosols in the Human Respiratory Tract* (ed. W. Hoffmann), pp. 34–40. Salzburg, Facultas.

Evans, R.D. (1967) On the carcinogenicity of inhaled radon decay products in man. In: *Radiation exposure of uranium miners*, Hearings before Joint Committee on Atomic Energy, US Government Printing Office, Washington, D.C.

Feely, H.W., Toonkel, L. & Larsen, R. (1981) Radionuclides and trace elements in surface air. Report EML 395, Appendix. New York, U.S. Department of Energy.

Fontan, J., Blanc, D., Huertas, M.L. & Marty, A.M. (1969) Measure de la mobilité et du coefficient de diffusion des particules radioactives. In: *Planetary Electrodynamics*, ed. C. Coroniti & J. Hughes, Vol. I, pp. 257–67, New York, Gordon & Breach.

Frey, G., Hopke, P.K. & Stukel, J.J. (1981) Effect of trace gases and water vapour on the diffusion coefficient of polonium-218. *Science*, **211**, 480–1.

Fry, L.M. & Menon, K.K. (1962) Determination of the troposphere residence time of lead-210. *Science*, **137**, 994–5.

Fuchs, N.A. (1959) *Evaporation and Droplet Growth in Gaseous Media*, Oxford, Pergamon.

Gavini, M.B., Beck, J.N. & Kuroda, P.K. (1974) Mean residence times of the long-lived radon daughters in the atmosphere. *Journal of Geophysical Research*, **79**, 4447–52.

George, A.C. (1972) Indoor and outdoor measurements of natural radon and radon decay products in New York City air. In: *The Natural Radiation Environment II*, ed. J.A.S. Adams, W.M. Lowder & T.F. Gesell, CONF-720805, pp. 741–50, Springfield, Va. NTIS.

 (1984) Passive integrated measurement of indoor radon using activated carbon. *Health Physics*, **46**, 867–72.

George, A.C. & Breslin, A.J. (1969) Deposition of radon daughters in humans exposed to uranium mine atmospheres. *Health Physics*, **17**, 115–24.

George, A.C. & Hinchliffe, L. (1972) Measurements of uncombined radon daughters in uranium mines. *Health Physics*, **23**, 791–803.

George, A.C., Hinchliffe, L. & Sladowski, R. (1975) Size distribution of radon daughter products in uranium mine atmospheres. *American Industrial Hygiene Journal*, **34**, 484–90.

Gesell, T.F. (1983) Background atmosphere ^{222}Rn concentrations outdoors and indoors: a review. *Health Physics*, **45**, 289–302.

Gormley, P.G. & Kennedy, M. (1949) Diffusion from a stream flowing through a cylindrical tube. *Proceedings of the Royal Irish Academy*, **52**, 163–9.

Graustein, W.C. & Turekian, K.K. (1986) ^{210}Pb and ^{127}Cs in air and soils measure the rate and vertical profile of aerosol scavenging. *Journal of Geophysical Research*, **91**, 14355–66.

Hess, C.T., Korsak, J.K. & Einloth, C.J. (1987) Radon in houses due to radon in potable water. In: *Radon and its Decay Products – Occurrence, Properties, and Health Effects*, ed. P.K. Hopke, Washington, American Chemical Society, pp. 30–41.

Huertas, M.L., Marty, A.M., Fontan, J., Alet, I. & Duffa, G. (1971) Measurement of mobility and mass of atmospheric ions. *Aerosol Science*, **2**, 145–50.

Hultqvist, B. (1956) Studies on naturally occurring ionising radiations. *Kungl. Svenska Vetenskapsakademiens Handlingar, Stockholm* **6**, No. 3.

Hursh, J.B., Schraub, A., Sattler, E. & Hoffmann, H. (1969) Fate of ^{212}Pb inhaled by human subjects. *Health Physics*, **16**, 257–67.

Ingersoll, J.G. (1983) A survey of radionuclide contents and radon emanation rates in building materials used in the U.S. *Health Physics*, **45**, 363–8.

International Commission on Radiological Protection (1981) Statement and recommendations of the International Commission on Radiological Protection from its 1980 meeting. ICRP Publication 26, reprinted with additions. Oxford. Pergamon.

(1986) Radiation protection of workers in mines. *Annals of the ICRP*, **16**, No. 1.

(1988) Lung cancer risk from environmental exposure to radon daughters. *Annals of the ICRP*, **17**, No. 1.

Israel, H. (1951) Radioactivity of the atmosphere. In: *Compendium of Meteorology*, ed. T.F. Malone, pp. 155–61, Boston, American Meteorological Society.

Israel, G.W. (1965) Thoron (Rn-220) measurements in the atmosphere and their application to meteorology. *Tellus*, **17**, 383–8.

Israel, H. & Israel, G.W. (1965) A new method of continuous measurement of radon (Rn-222) and thoron (Rn-220) in the atmosphere. *Tellus*, **18**, 557–61.

Jacobi, W. (1972) Activity and potential alpha energy of ^{222}radon and ^{220}radon daughters in different air atmospheres. *Health Physics*, **22**, 441–50.

Jacobi, W. & André, K. (1963) The vertical distribution of radon 222 and radon 220 and their decay products in the atmosphere. *Journal of Geophysical Research*, **68**, 3799–814.

Jacobi, W. & Paretzke, H.G. (1985) Risk assessment for indoor exposure to radon daughters. *The Science of the Total Environment*, **45**, 551–62.

James, A.C. (1987a) A re-consideration of cells at risk and other key factors in radon dosimetry. In: *Radon and its Decay Products, Properties and Health Effects*, ed. P.K. Hopke, Washington D.C. American Chemical Society, pp. 400–18.

(1987b) Lung dosimetry for radon and thoron daughters. In: *Radon and its Progeny in Indoor Air*, ed. W.W. Nazaroff & A.V. Nero, New York, Wiley, pp. 259–309.

Jonassen, N. & Hayes, E. (1972) Mobility distribution of radon-222 daughter small ions in laboratory air. *Journal of Geophysical Research*, **77**, 5876–82.

Jost, W. (1960) *Diffusion in Solids, Liquids and Gases*, New York, Academic Press.

Junge, C.E. (1963) *Air Chemistry and Radioactivity*, New York, Academic Press.

Keller, G., Folkerts, K.H. & Muth, H. (1982) Methods for the determination of ^{222}Rn (radon) and ^{220}Rn (thoron) exhalation rates using alpha spectroscopy. *Radiation Protection Dosimetry*, **3**, 83–9.

Keller, G. & Folkerts, K.H. (1984) Radon-222 concentrations and decay product equilibrium in dwellings and in the open air. *Health Physics*, **47**, 385–98.

Knutson, E.O., George, A.C., Frey, J.J. & Koh, B.R. (1983) Radon daughter plate out, II Prediction model. *Health Physics*, **45**, 445–51.

Kotrappa, P., Bhanti, D.P., Menon, V.B., Dhaudayutham, R., Gohel, C.O. & Nambiar, P.P. (1976) Assessment of airborne hazards in the thorium processing industry. *American Industrial Hygiene Journal*, **37**, 613–16.

Kraner, A.W., Schroeder, G.L. & Evans, R.D. (1964) Measurements of the effects of atmospheric variables on radon-222 flux and soil gas concentrations. In: *The Natural Radiation Environment*, ed. J.A.S. Adams & W.M. Lowder, University of Chicago Press, pp. 191–215.

Krzesniak, J.W. & Porstendörfer, J. (1978) Diffusion coefficients of airborne radio-active iodine and methyl iodide. *Health Physics*, **35**, 417–31.

Kruger, J. & Andrews, M. (1976) Measurements of the attachment coefficient of radon-220 decay products to monodispersed polystyrene aerosols. *Journal of Aerosol Science*, **7**, 21–36.

Kruger, J. and Nothling, J.F. (1979) A comparison of the attachment of the decay products of radon-220 and radon-222 to monodispersed aerosols. *Journal of Aerosol Science*, **10**, 571–9.

Lambert, G., Ardouin, B. & Polian, G. (1982) Volcanic output of long lived radon daughters. *Journal of Geophysical Research*, **87**, 1103–8.

Lehmann, L. & Sittkus, A. (1959) Bestimmung von Aerosolverweilzeiten aus dem RaD und RaF Gehalt der atmosphärischen Luft und die Niederschlage. *Naturwissenschaften*, **46**, 9–10.

Lockhart, L. (1960). Atmospheric radioactivity in South America and Antarctica. *Journal of Geophysical Research*, **65**, 3999–4005.

Lucas, H.F. (1957) Improved low-level alpha-scintillation counts for radon. *Review of Scientific Instruments*, **28**, 680–3.

  (1964) A fast and accurate survey technique for both radon 222 and radon 220. In: *The Natural Radiation Environment*, ed. J.A.S. Adams & W.M. Lowder, University of Chicago Press, pp. 315–29.

Magno, P.J., Groulx, P.R. & Apidianakis, J.C. (1970) Lead-210 in air and in total diets in the United States during 1966. *Health Physics*, **18**, 383–8.

Machta, L. & Lucas, H.F. (1962) Radon in the upper atmosphere. *Science*, **135**, 296–9.

Megaw, W.J. & Wiffen, R.D. (1961) The generation of condensation nuclei by ionising radiation. *Geofisica Pura et Applicata*, **50**, 118–26.

Mercer, T.T. (1976) The effect of particle size on the escape of recoiling RaB atoms from particulate surfaces. *Health Physics*, **31**, 173–5.

Mercer, T.T. & Stowe, W.A. (1969) Deposition of unattached radon decay products in an impactor stage. *Health Physics*, **17**, 259–64.

Moore, H.E., Poct, S.E. & Martell, E.A. (1972) Tropospheric residence times indicated by radon and radon daughter concentrations. In: *Natural Radiation Environment II*, CONF-720805, ed. J.A.S. Adams, W.M. Lowder & T.F. Gesell, Springfield, Va, pp. 775–85.

  (1973) ^{222}Rn, ^{210}Pb, ^{210}Bi and ^{210}Po profiles and aerosol residence times versus altitude. *Journal of Geophysical Research*, **78**, 7065–75.

Munson, R.J., Tyndall, A.M. & Hoselitz, K. (1939) The mobility of alkali ions in gases. *Proceedings of the Royal Society A*, **172**, 28–54.

Myrick, T.E., Berven, B.A. & Haywood, F.F. (1983) Determination of concentrations of selected radionuclides in surface soils in the U.S. *Health Physics*, **45**, 631.

National Council on Radiological Protection and Measurements (1984) Evaluation of occupational and environmental exposures to radon and radon daughters in the United States. NCRP Report 78, Bethesda, Md.

  (1988) Measurement of radon and radon daughters in air. NCRP Report 97. Bethesda, Md.

National Radiological Protection Board (1990). Human exposure to radon in homes. *Documents of the NRPB*, **1**, 17–32.

National Research Council, Committee on the Biological Effects of Ionising Radiations (1988) Health risks of radon and other internally deposited alpha emitters (BEIR IV). National Academy Press. Washington, DC.

Nero, A.V. (1983) Airborne radionuclides and radiation in buildings: a review. *Health Physics*, **45**, 303–22.

  (1988) Estimated risk of lung cancer from exposure to radon decay products in U.S. homes: a brief review. *Atmospheric Environment*, **22**, 2205–11.

Nero, A.V. & Nazaroff, W.W. (1984) Characterizing the source of radon indoors. *Radiation Protection Dosimetry*, **7**, 23–9.

Nero, A.V., Schwehr, M.B., Nazaroff, W.W. & Revzan, K.L. (1986) Distribution of airborne radon-222 concentrations in U.S. homes. *Science*, **234**, 992–7.

O'Riordan, M.C., James, A.C. & Brown, K. (1982) Some aspects of human exposure to ^{222}Rn decay products. *Radiation Protection Dosimetry*, **3**, 75–82.

O'Riordan, M.C., James, A.C., Green, B.M.R. & Wrixon, A.D. (1987) Exposure to radon daughters in dwellings. Report NRPB-GS6, Chilton, Oxon, National Radiological Protection Board.

Pearson, J.E. & Jones, G.E. (1966) Soil concentrations of emanating radium-226 and the emanation of radon-222 from soils and plants. *Tellus*, **18**, 655–61.

Peirson, D.H., Cambray, R.S. & Spicer, G.S. (1966) Lead-210 and polonium-210 in the atmosphere. *Tellus*, **18**, 427–33.

Porstendörfer, J. (1984) Behaviour of radon daughter products in indoor air. *Radiation Protection Dosimetry*, **7**, 104–14.

Porstendörfer, J. & Mercer, T.T. (1978) Adsorption probability of atoms and ions on particulate surfaces in submicrometer size range. *Journal of Aerosol Science*, **9**, 469–74.

   (1979) Influence of electric charge and humidity upon the diffusion coefficient of radon decay products. *Health Physics*, **37**, 191–9.

Porstendörfer, J., Röbig, G. & Ahmed, A. (1979) Experimental determination of the attachment coefficients of atoms and ions on monodisperse aerosols. *Journal of Aerosol Science*, **10**, 21–8.

Porstendörfer, J., Reineking, A. & Becker, K.H. (1987) Free fractions, attachment rates and plate-out rates of radon daughters in houses. In: *Radon and its Decay Products*, ed. P.K. Hopke, Washington D.C., American Chemical Society, pp. 285–300.

Put, L.W. & de Meijer, R.J. (1985) Survey of radon concentrations in Dutch dwellings. *The Science of the Total Environment*, **45**, 389–96.

Raabe, O.G. (1968) Measurement of the diffusion coefficient of radium A. *Nature*, **217**, 1143–5.

Raes, F. (1985) Description of the properties of unattached ^{218}Po and ^{212}Pb particles by means of the classical theory of cluster formation. *Health Physics*, **49**, 1177–87.

Raes, F., Janssens, A. & Vanmarcke, H. (1985) A closer look at the behaviour of radioactive decay products in air. *The Science of the Total Environment*, **45**, 205–18.

Raghunath, B. & Kotrappa, P. (1979) Diffusion coefficients of decay products of radon and thoron. *Journal of Aerosol Science*, **10**, 133–8.

Reineking, A., Becker, K.M. & Porstendörfer, J. (1985) Measurements of the unattached fractions of radon daughters in houses. *The Science of the Total Environment*, **45**, 261–70.

Rutherford, E. & Brooks, H.T. (1901) The new gas from radium. *Transactions of Royal Society of Canada*, **7**, 21–5.

Satterly, J. (1908) The amount of radium emanation in the atmosphere. *Philosophical Magazine*, **16**, 584–615.

Schery, S.D. (1985) Measurements of airborne ^{212}Pb and ^{220}Rn at varied indoor locations within the United States. *Health Physics*, **49**, 1061–7.

Schery, S.D., Gaeddert, D.H. & Wilkening, M.H. (1980) Two-filter monitor for atmospheric ^{222}Rn. *Review of Scientific Instruments* **51**, 338–43.

   (1984) Factors affecting exhalation of radon from a gravelly sandy loam. *Journal of Geophysical Research*, **89**, 7299–309.

Schiller, C., Gebhart, J., Heyder, J., Rudolf, G. & Stahlhofen, W. (1988) Deposition of monodisperse insoluble aerosol particles in the 0.005 to 0.2 μm size range within the human respiratory tract. *Annals of Occupational Hygiene*, **32**, 41–9.

Schmier, H. & Wicke, A. (1985) Results from a survey of indoor radon exposure in the Federal Republic of Germany. *The Science of the Total Environment*, **45**, 307–10.

Shahin, M.M. (1966) Mass-spectrometric studies of corona discharges in air at atmospheric pressure. *Journal of Chemical Physics*, **45**, 2600–5.

Sinnaeve, J., Clemente, G. & O'Riordan, M. (1984) The emergence of natural radiation. *Radiation Protection Dosimetry*, **7**, 15–17.

Standen, E., Kilstad, A.K. & Lind, B. (1984) The influence of moisture and temperature on radon exhalation. *Radiation Protection Dosimetry*, **7**, 55–8.

Stewart, N.G. & Crooks, R.N. (1958) Long range travel of the radioactive cloud from the accident at Windscale. *Nature*, **182**, 627–30.

Swedjemark, G.A. & Mjönes, L. (1984) Radon and radon daughter concentrations in Swedish houses. *Radiation Protection Dosimetry*, **7**, 341–6.

Talbot, R.W. & Andren, A.W. (1983) Relationship between Pb and ^{210}Pb in aerosol and precipitation at a semi-remote site in northern Wisconsin. *Journal of Geophysical Research*, **88**, 6752–60.

Tanner, A.B. (1964) Radon migration in the ground: a review. In: *The Natural Radiation Environment*, ed. J.A.S. Adams & W.M. Lowder, University of Chicago Press, pp. 161–90.

  (1980) Radon migration in the ground: a supplementary review. In: *The Natural Radiation Environment III*, ed. T.F. Gesell & W.M. Lowder, Springfield, Va., pp. 5–56.

Thomas, J.W. & LeClare, P. (1970) A study of the two filter method for radon-222. *Health Physics*, **18**, 113–22.

Tu, K.W. & Knutson, E.O. (1984) Total deposition of ultrafine hydrophobic and hygroscopic aerosols in the human respiratory system. *Aerosol Science & Technology*, **3**, 453–65.

Turekian, K.K., Nozaki, Y. & Benninger, L.K. (1977) Geochemistry of atmospheric radon and radon products. *Annual Review of Earth Planetary Science*, **5**, 227–55.

Urban, M. & Piesch, E. (1981) Low level environmental radon dosimetry with a passive track etch detector device. *Radiation Protection Dosimetry*, **1**, 97–109.

United Nations Scientific Committee on Effects of Atomic Radiations (1982) Ionising radiation: sources and biological effects. New York, United Nations.

United States Public Health Service (1957) Control of radon and daughters in uranium mines and calculations on biological effects. Washington D.C.

Vanmarcke, H., Janssens, A., Raes, F., Poffijn, A., Berkens, P. & Van Dingenen, R. (1987) The behaviour of radon daughters in the domestic environment: effect on the effective dose equivalent. In: *Radon and its Decay Products*, ed. P.K. Hopke, Washington, D.C., American Chemical Society, pp. 301–23.

Watnick, S., Latner, N. & Graveson, R.T. (1986) A ^{222}Rn monitor using α spectroscopy. *Health Physics*, **50**, 645–6.

Whitby, K.T. (1978) The physical characteristics of sulfur aerosols. *Atmospheric Environment*, **12**, 135–59.

Wilkening, M. (1987) Effect of radon on some electrical properties of indoor air. In: *Radon and its Decay Products*, ed. P.K. Hopke, Washington D.C., American Chemical Society, pp. 377–97.

Wilkening, M.H., Clements, W.E. & Stanley, D. (1972) Radon-222 flux measurements in widely separated regions. In: *Natural Radiation Environment II*, ed. J.A.S. Adams, W.M. Lowder & T.F. Gesell, pp. 717–30, Springfield, Va., NTIS.

Wilkening, M.H., Kawano, M. & Lane, C. (1966) Radon-daughter ions and their relation to some electrical properties of the atmosphere. *Tellus*, **18**, 679–684.

Wilkening, M. & Romero, V. (1981) ^{222}Rn and atmospheric electrical parameters in the Carlsbad caverns. *Journal of Geophysical Research*, **86**, 9911–16.

Wrixon, A.D. *et al.* (1988) Natural radiation exposure in UK dwellings, Report NRPB-R190. Chilton, Oxon., National Radiological Protection Board.

Yu, C.P. & Diu, C.K. (1982) A comparative study of aerosol deposition in different lung models. *American Industrial Hygiene Journal*, **43**, 54–65.

Zarcone, M.J., Schery, S.D., Wilkening, M.H. & McNamee, E. (1986) A comparison of measurements of thoron, radon and their daughters in a test house with model predictions. *Atmospheric Environment*, **20**, 1273–9.

2

○ ○ ○ ○ ○ ○ ○ ○ ○ ○ ○ ○ ○ ○ ○ ○ ○ ○ ○

Fission product aerosols

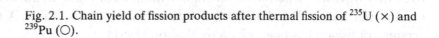

2.1 Process of fission

When a uranium or plutonium nucleus undergoes fission, it splits into two fragments of unequal size. Figure 2.1 shows yield/mass curves for fission of ^{235}U and ^{239}Pu by slow neutrons. Fission by fast neutrons gives slightly different distributions, with increased prob-

Fig. 2.1. Chain yield of fission products after thermal fission of ^{235}U ($\times$) and ^{239}Pu ($\bigcirc$).

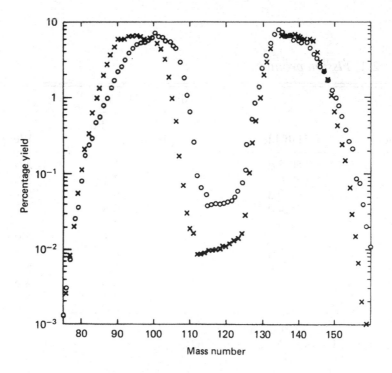

ability of nearly symmetrical fission, so that the dip in the middle of the curve is less pronounced.

An example of a possible mode of fission is

$$^{235}U_{92} + n \rightarrow {}^{90}Kr_{36} + {}^{144}Ba_{56} + 2n$$

The prompt neutrons emitted in fission are available for fission in other nuclei – hence the chain reaction. The fission fragments formed initially are rich in neutrons. For example the heaviest stable isotopes of krypton and barium are ^{86}Kr and ^{138}Ba. Excess neutrons are emitted from the fission fragments as delayed neutrons or converted to protons by beta decays. For example

$$^{90}Kr_{36} \rightarrow {}^{90}Rb_{37} \rightarrow {}^{90}Sr_{38} \rightarrow {}^{90}Y_{39} \rightarrow \text{Stable Zr}$$
$$\text{33 s} \qquad \text{2.7 m} \qquad \text{28 a} \qquad \text{64 h}$$

In general, the radioactive half-life increases down the chain of fission products, but some long-lived fission products, for example ^{90}Sr, have short-lived daughters.

Altogether, there are about 200 radioactive fission products with half-lives varying from a fraction of a second to millions of years. Some of the most important are listed in Table 2.1 with their yields in fission. A full account of fission has been given by Walton (1961).

Table 2.1. *Fission products*

Nuclide	Half-life	Fission yield (%) ^{235}U (thermal)	^{239}Pu	Weapons
^{89}Sr	50.5 d	4.8	1.7	2.6
^{90}Sr	28.6 a	5.9	2.1	3.5
^{95}Zr	64.0 d	6.5	5.0	5.1
^{103}Ru	39.4 d	3.2	6.8	5.2
^{106}Ru	1.0 a	0.4	4.3	2.4
^{131}I	8.0 d	2.8	3.8	2.9
^{137}Cs	30.2 a	6.2	6.6	5.6
^{140}Ba	12.8 d	6.3	5.6	5.2
^{141}Ce	32.5 d	5.8	5.3	4.6
^{144}Ce	284 d	5.4	3.8	4.7

Thermal yields from Crouch (1977).
Weapons yields from UNSCEAR (1962).

The activities of each fission product for a given amount of fission energy can be calculated, but there may be uncertainties in the contributions of slow and fast fission and in the relative contributions of fission in ^{235}U, ^{238}U and ^{239}Pu. The energy released in nuclear explosions is reported in kilotons (kT) of conventional chemical high explosive equivalent. One kT is taken as equivalent to 10^{12} calories or 4.2×10^{19} erg. The prompt release of energy per fission is 180 MeV or 2.9×10^{-4} erg, so 1 kT is equivalent to 1.45×10^{23} fissions. If the yield or fission probability of a particular fission product is f, and its decay constant is $\lambda \, \text{s}^{-1}$, its activity following a 1-kT nuclear explosion is $1.45 \times 10^{23} \, f\lambda$ Bq. For example, for ^{137}Cs, $f = 0.056$, $\lambda = 7.3 \times 10^{-10} \, \text{s}^{-1}$ and the activity is 5.9×10^{12} Bq, or 5.9 TBq per kT.

In a reactor, the energy per fission, including the energy of the delayed neutrons and of the fission products, is 200 MeV. To produce 1 MW thermal energy, 3.1×10^{16} fissions per second are required. If the half-life of the fission product is short compared with the duration of operation of the reactor, its activity comes into equilibrium when creation by fission equals radioactive decay. Assuming a constant level of power for a duration of T secs, the activity is $3.1 \times 10^4 f(1 - \exp{-\lambda T})$ TBq per MW. Some fission products themselves absorb neutrons (the socalled reactor poisons) and for them the calculation of activity is more complicated. Figure 2.2 shows the combined activity of 1 g of fission products formed in an instantaneous burst of fission and also from 1 g of fission products formed over a period of a year (Walton, 1961). The activity from a short burst decays approximately as $t^{-1.2}$.

Table 2.2 shows the activity of four of the most important fission products per kT nuclear explosion and per MW reactor fuel. The activity of long-lived fission products is approximately proportional to the megawatt-days (MWd) of operation, and in Table 2.2 one year's operation has been assumed. Some nuclides, including ^{90}Sr, have quite different yields in slow neutron fission of ^{235}U and ^{239}Pu. At high burn-up, fission in ^{239}Pu (formed by neutron capture in ^{238}U and subsequent decay of ^{239}U) contributes to the power. Consequently, the ratio of activity ^{90}Sr/^{137}Cs depends on the burn-up.

^{133}Cs is a stable isotope, formed in 6% of fissions. Neutron capture in ^{133}Cs gives ^{134}Cs, which is radioactive with half-life 2.06 a. In the Chernobyl reactor, after 10,000 MWd-Te irradiation, the ratio ^{134}Cs/^{137}Cs was about 0.5. By measuring the ratio in soil, Chernobyl ^{137}Cs could be distinguished from ^{137}Cs deposited previously due to weapon tests.

2.2   Major releases of fission products

Table 2.3 shows the estimated releases of ^{90}Sr, ^{131}I, ^{137}Cs and ^{144}Ce in the Nevada tests, the thermonuclear tests (H tests), the 1957 Windscale accident, the 1957 accident at a separation plant in the Urals

Fig. 2.2. Activity of 1 g of fission products produced instantaneously (A) and over 1 a (B). Dashed line is $t^{-1.2}$.

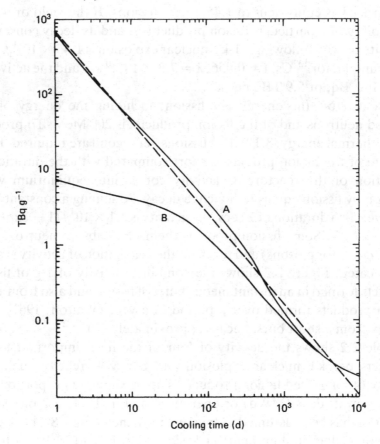

Table 2.2. *Fission product activity*

Nuclide	TBq per kT weapon yield	TBq per MW reactor power after 1 a
^{90}Sr	4	50
^{131}I	4000	900
^{137}Cs	6	45
^{144}Ce	200	900

and the 1986 Chernobyl accident. The release of ^{131}I from Chernobyl was only a fifth of the release in the Nevada tests. It was less than 1/1000 of the release in H tests, but most of the ^{131}I from the latter decayed in the stratosphere and only a small fraction reached ground. The effects of the release of ^{131}I are considered in Chapter 3.

The ratio of activities ^{137}Cs/^{131}I is much higher in reactor than in bomb fission products. The release of ^{137}Cs from Chernobyl was 10 times greater than that from the Nevada tests, though only about 4% of the release in H tests.

2.3   Particle size and chemical fractionation

The particle size of a fission aerosol, and the distribution of fission products between particulate and vapour phases, depends on the mechanism of release to the atmosphere. In a weapons explosion, some physicochemical fractionation of radionuclides may occur, particularly if the explosion is near the ground. Everything in the vicinity is vapourised by the heat of the explosion, but within less than a minute the fireball cools to a temperature in the range 1000–2000°C, and refractory materials such as metal oxides and silicates condense to form particles (Glasstone & Dolan, 1977). Refractory fission products, and plutonium, are incorporated in these particles.

Less-refractory fission products condense later onto the surface of the particles. Those with gaseous precursors, for example ^{90}Sr and ^{137}Cs, condense as they are formed by decay of their parent nuclides. The

Table 2.3. *Release of fission products in weapons tests and accidents*

Source	^{90}Sr (TBq)	^{131}I (TBq)	^{137}Cs (TBq)	^{144}Ce TBq)
Nevada atmospheric tests (total yield 1 MT)	4×10^3	4×10^6	6×10^3	2×10^5
All H-tests 200 MT)	8×10^5	8×10^8	1×10^6	4×10^7
Windscale accident	0.3	9×10^2	80	~10
Chemical explosion in Urals	4×10^3	—	30	5×10^4
Three Mile Island accident	—	0.6	—	—
Chernobyl accident	8×10^3	6×10^5	4×10^4	9×10^4

refractory fission products are distributed between particles according to their mass, or diameter cubed. The more volatile fission products, and those with volatile parents, are distributed according to a power of d_p between 1 and 2 (compare equation 1.18). Per unit mass, there is then more activity in the smaller particles. The particle size depends on the volumetric concentration of condensing material, and decreases as the fireball expands. The earlier a fission product condenses, the larger the particles in which it is incorporated, and the quicker it is lost by fallout to the ground.

At distances of about 100 km from ground zero of the Nevada tests, the particle size of fallout ranged from a few μm to a few 100 μm. Fission products with volatile precursors were enhanced by about a factor two compared with refractory fission products. The fractionation was greater the smaller the particles (Hicks, 1982).

The fireballs created when H-bombs are exploded at high altitude are very large, but no ground-based material is incorporated, so the condensation aerosol particles are very small, with 90% of the activity in particles less than 0.3 μm in diameter (Drevinsky & Pecci, 1965). Coagulation takes place with the natural stratospheric aerosol, which is thus labelled with fission products. In the troposphere there is further coagulation, and near ground most activity is in the 0.3–1-μm size range (Lockhart et al., 1965). Any fractionation of fission products in H-test explosions is evened out by the process of coagulation, and there is no evidence of fractionation in particles sampled near ground.

2.4   Emission from reactors

It is convenient to consider reactor accidents alongside weapon explosions so that the release of fission products can be compared, but the mode of dispersion is quite different. The configuration and thermal capacity of power reactors are such that bomb-like explosions are not possible. In the Chernobyl accident, nuclear overheating, a steam explosion and steam/zirconium reactions all contributed to the disruption of the reactor (U.S.S.R. State Committee, 1986), but the long-distance environmental effects were due to the subsequent releases of fission products from the damaged reactor.

Reactor fuel elements are contained in cans. In early British reactors, these were made of aluminium or aluminium/magnesium alloy to minimise capture of neutrons in the canning material. Nowadays, uranium fuel is enriched with respect to the ^{235}U content, and the extra reactivity enables steel or zirconium cans to be used. In the original

Windscale reactors, and some US reactors of the 1950 period, the fuel was cooled by air blown straight to atmosphere, and no use was made of the heat to produce power. In all power reactors now operating, the coolant is contained in a closed-circuit pressure vessel. Outer containment buildings, which can also withstand some pressure in the event of failure or leakage from the pressure circuit, enclose the US pattern pressurised water and boiling water reactors, but no such provision was made for the Russian boiling water reactor at Chernobyl. All defences (cans, pressure vessel, containment building if provided) must be breached before fission products can be released to atmosphere.

Can failures occur from time to time. The release of fission products from them depends on the temperature and type of fuel. If the fuel is uranium metal, as in the Windscale and Magnox reactors, and the can fails, the uranium will oxidise in air or CO_2. In laboratory experiments, the mass median aerodynamic equivalent diameter (*MMAD*) of the particles produced by oxidation of uranium increased from about 40 μm when the temperature of oxidation was 600°C to 500 μm at 1000°C (Megaw *et al.*, 1961). At high temperature, a coherent sintered oxide layer formed on the uranium and this hindered the formation of particles.

When oxidation was in air, a substantial fraction of the ^{131}I in the oxidised uranium was released as vapour, but in CO_2 only a small percentage appeared in the gas phase, Apart from iodine and the noble gases, other fission products were found in the oxide particles with the same activity per gram of uranium as in the fuel element. On melting metallic uranium in air or CO_2, Megaw *et al.* found that 5–15% of ^{131}I was released, but only 0.2% on melting in argon. Oxidation is required to dissociate the uranium iodide which is probably the chemical form of ^{131}I in fuel. Megaw *et al.* worked with trace-irradiated uranium, and did not examine the release of ^{137}Cs, which was at too low a level of activity.

In Advanced Gas Cooled (AGR), Pressurised Water (PWR) and Boiling Water (BWR) reactors, and in the Russian RMBK, the fuel is UO_2. Experiments in the UK and USA, reviewed by Farmer & Beattie (1976), showed less than 1% release of fission product iodine and caesium from punctured UO_2 fuel cans at about 1000°C in air or steam, rising to 10–50% release at 1800°C. At 2800°C, the UO_2 melted and there was nearly complete release of volatile nuclides (I, Te, Cs, Ru) but only small release of refractory alkaline earth and rare earth nuclides.

In more recent work at Oak Ridge National Laboratory, Osborne *et al.* (1986) heated specimens of irradiated UO_2 fuel in steam atmospheres. As ^{131}I had decayed to negligible levels between irradiation and test, the long-lived [129]I was used to estimate release of iodine. The fractional releases of [85]Kr, [129]I and ^{137}Cs are plotted against inverse of absolute temperature in Fig. 2.3. The results were fairly consistent with the calculations of the US Nuclear Regulatory Commission (1981). Only slight differences were found in the release of the three nuclides.

Thermodynamic calculations indicate that Cs and I should be released from PWR fuel as CsI. However radiolysis may decompose CsI and volatile iodine compounds may be created, particularly if the pH of the coolant water is not high (Postma & Pasedag, 1986).

Some historical instances of radioactive emissions will now be reviewed.

Fig. 2.3. Release of isotopes Kr (○), I (△) and Cs (□) from fuel specimens in experiments at Oak Ridge. Line is calculation by U.S. Nuclear Regulatory Commission (1981).

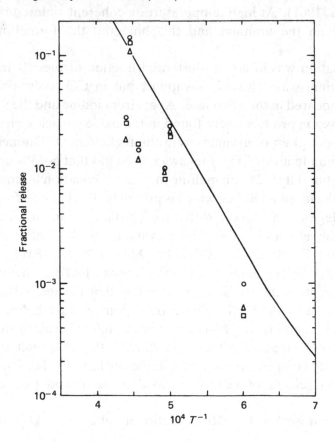

2.5   Windscale emissions, 1952–6

The two Windscale piles were fuelled with natural uranium canned in aluminium. Coolant air was blown through the reactor and exhausted from a 120-m stack (Fig. 2.4). Filters were installed at the top of the stack, but were not very effective. Some fuel cans developed pinholes during operation, and others became damaged and lodged in the ducts behind the pile. It is estimated that about 20 kg of irradiated uranium were disseminated to atmosphere as oxide particles from these cans (Stather *et al.*, 1986). The temperature of oxidation was 200–400°C. The particle size, measured at the top of the stack, showed a mass median diameter of 35 μm (Mossop, 1960).

The fallout pattern of the oxide particles in the vicinity of the piles was assessed in 1955–6 by Geiger counter surveys of the beta-dose rate near the ground, and subsequently, in 1958/61 by measurements of ^{90}Sr and ^{137}Cs in soil (Ellis *et al.*, 1960; Booker, 1962; Chamberlain, 1987). The ^{90}Sr/^{137}Cs ratio in the oxide particles was near unity, there being little fractionation by volatilisation at the low temperatures of oxidation.

In Fig. 2.5 the fallout of the ^{137}Cs in the pre-1957 oxide particles, is compared with the fallout in the Windscale accident of October 1957.

Fig. 2.4. Windscale pile.

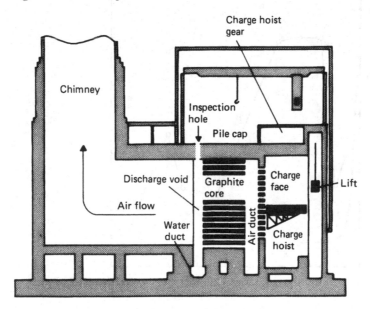

The different pattern of fallout is mainly related to the particle size of the aerosol in the circumstances. The cumulative fallout of ^{137}Cs in Cumbria from distant weapon tests reached a peak of 4 to 7 kBq m^{-2} (depending on annual rainfall) in 1964. Thus a few farms within about 1 km of the Windscale site received two to four times as much fallout of ^{137}Cs and ^{90}Sr from the oxide particles as they did subsequently from weapons tests. At the nearest large settlement, Seascale, which is 3 km from the Windscale site, the fallout in oxide particles and in bomb debris were of similar magnitude.

Few measurements of activity in milk or other foods were made during the period (1954–6) when most of the emissions occurred. Measurements in 1958/9 (Ellis *et al.*, 1960; Chamberlain, 1987) showed 0.5 to 5 Bq ^{90}Sr per g Ca in milk, considerably higher than the contemporary concentrations elsewhere in the UK, but similar to the range found in 1962/5 at the peak of the bomb fallout.

Dissemination of oxide particles from certain early types of reactor may also have occurred in the USA in the late 1940s, but details have not been published. There are now no air-cooled reactors in operation, but dissemination of oxide particles could happen if there were an

Fig. 2.5. ^{137}Cs in soil near Windscale: A, pre-1957 oxide particles; B, accident of Oct. 1957; F, weapons test fallout.

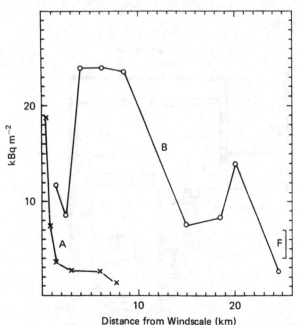

accident during handling of uranium metal fuel. The early Windscale emissions are of historical interest as the only documented example of fission products released in the proportions found in fuel.

2.6   Windscale accident of October 1957

During a planned release of Wigner energy from graphite in Windscale No. 1 Pile, it became overheated. Oxidation of the graphite raised the temperature further, despite attempts to restrict access of air, and part of the reactor core reached an estimated temperature of 1300°C (Penney, 1957). About 6 to 8 tonne of uranium melted, but, in contrast to the previous operational experience, there was remarkably little dissemination of particulate uranium oxide (Chamberlain & Dunster, 1958; Chamberlain, 1981). The high temperature and restricted air flow probably caused a skin of sintered oxide to form on the uranium.

Table 2.4 shows the activities of fission products, relative to ^{137}Cs, in the reactor fuel, the stack filter and sampling filters in the environment. All the environmental samples showed activity of the relatively volatile fission products, isotopes of Te, I and Cs, and only traces of the refractory elements, such as Zr, the rare earths and the alkaline earths.

Referring to the stack filter, the report of the Court of Inquiry (Penney, 1957) stated: 'Iodine vapour had come through the filter but the major part of the particulate material had been caught by the filter'. This was published before analyses of the filter material were available, and was incorrect. Most of the activity on the filter was associated with a yellow powder (Crouch & Swainbank, 1958). The powder comprised chain aggregates of submicrometre particles of lead and bismuth oxides, which originated as a fume in the reactor. The distribution of fission products on the powder, as between volatile and refractory elements, was similar to that found in the environment.

The total activity on the stack filter is not known accurately (Chamberlain, 1981), but its collection efficiency was not high, which is not surprising since it was not designed for the conditions which prevailed in the accident. The filter was certainly not responsible for the separation of volatile from refractory fission products. Most of the latter, together with the uranium and plutonium, remained in the fuel channels.

The ratio of activities ^{131}I/^{137}Cs on 24 environmental filters from various parts of the UK averaged $9.1 \pm$ (S.D.)0.9 (Stewart *et al.*, 1961). This is less than half the ratio in the reactor fuel, allowing for the fact that the reactor had been virtually shut down for 4 d before the accident. The paper filters used in 1957 for monitoring atmospheric pollution

Table 2.4. Activities relative to ^{137}Cs after Windscale accident

Nuclide	Reactor inventory[a]	Stack filter[b]	Environmental filters				
			Local (Calder)[c]	Harwell[d]	London[e]	All UK[f]	Europe[f]
^{137}Cs	1.00	1.00	1.00	1.00	1.00	1.00	1.00
^{133}Xe	41	—	—	—	150	—	
^{131}I	20	22	17	9	11	9.1	10 to 3
^{132}Te	20	—	—	8			
^{140}Ba	48	0.5	0.9	—			
^{89}Sr	35	0.3	0.4	0.07			
^{90}Sr	0.9	0.005	0.007	0.004			
^{95}Zr	54	0.08	0.1	—			
^{103}Ru	26	0.3	0.8	0.6			
^{141}Ce	48	0.07	—	—			
$^{239+240}$Pu	0.02	1.3×10^{-5}					
^{210}Po	—	0.18	0.14	0.24	0.22	0.23	0.26

(a) Reactor shut down for 4 d before accident. Irradiation 278 MWd/Te.
(b) Geometric mean of analyses of front and back of filter (Crouch & Swainbank, 1958).
(c) Chamberlain (1981).
(d) Stewart & Crooks (1958).
(e) Stewart et al. (1961) ^{133}Xe measurement on liquid air sample (Chamberlain, 1981).
(f) Stewart et al. (1961) ^{131}I/^{137}Cs ratio varied with distance of travel of cloud.

were operated at relatively low flow rates (about $30 \, l \, min^{-1}$) and were probably more efficient for gaseous radioiodine than modern high volume glass-fibre filters, but some penetration of ^{131}I must have occurred. The $^{131}I/^{137}Cs$ ratio declined with distance as the cloud from Windscale travelled over Europe, indicating preferential deposition of ^{131}I (Stewart & Crooks, 1958). It is unlikely, therefore, that CsI was the predominant form of iodine. Charcoal packs were not used to sample ^{131}I in air in 1957, and the proportions present as inorganic, or organic vapour, or adsorbed on particles, is not known.

Most of the caesium and other activities were released as fine particles, probably attached to metal fume, but some larger particles were found near the reactor (Chamberlain & Dunster, 1958). One 15-μm diameter particle appeared on gamma spectrometry to have only $^{140}Ba + {}^{140}La$ (3×10^4 Bq) but subsequent radiochemical analysis showed that it also had ^{89}Sr (1.5×10^4 Bq) and ^{90}Sr (3×10^3 Bq). It was believed that chromatographic separation of alkaline earth elements had occurred, by condensation on graphite, and that particles of graphite subsequently disseminated by combustion carried the activity. Particles with highly fractionated fission product activity were found after the Chernobyl release (Devell *et al.*, 1986).

Overall, however, the release of refractory fission products from Windscale was less than the release of volatile elements by two or three orders of magnitude, relative to the inventories in the reactor fuel (Table 2.4). Alpha activity on the stack filters and environmental filters was mainly ^{210}Po, derived from the bismuth irradiated in the isotope channels (Crouch & Swainbank, 1958; Crooks *et al.*, 1959). The $^{210}Po/^{137}Cs$ ratio on the environmental filters was about 0.2, with no significant change with distance, suggesting that both activities were carried on the same fume particles.

Examples of the dosages of ^{137}Cs deduced from the environmental filters (Stewart *et al.*, 1961; Crick & Linsley, 1982) are shown in Fig. 2.6. The dosage of activity in air is the integral over time of the volumetric concentration, with units $Bqs \, m^{-3}$ or $Bqd \, m^{-3}$, and is the activity in the sampler divided by the volumetric sampling rate. Measurements were made on filters from 24 locations in the UK, the nearest to Windscale being 60 to 90 km to the SE, in the industrial towns of Lancashire. The highest recorded dosages of ^{137}Cs and ^{131}I were 3.6 and 37 $Bqd \, m^{-3}$ respectively. By calculation (Clarke & Macdonald, 1978), it can be estimated that the dosages near the point of maximum concentration 5 km from Windscale were about 100 times greater.

The deposition of ^{137}Cs in the direction of maximum fallout SSE of Windscale is shown in Fig. 2.5. There was a maximum at 5 km downwind, where the plume from the 130-m stack reached the ground, and a second maximum at 20 km, where there is high ground. There was little rain during the passage of the plume. The ratio ^{131}I/^{137}Cs in deposited activity was 50 : 1 (Booker, 1962).

Various estimates of the release of fission products from Windscale

Fig. 2.6. Dosage of ^{137}Cs in air, Oct. 1957 (Bqd m^{-3}).

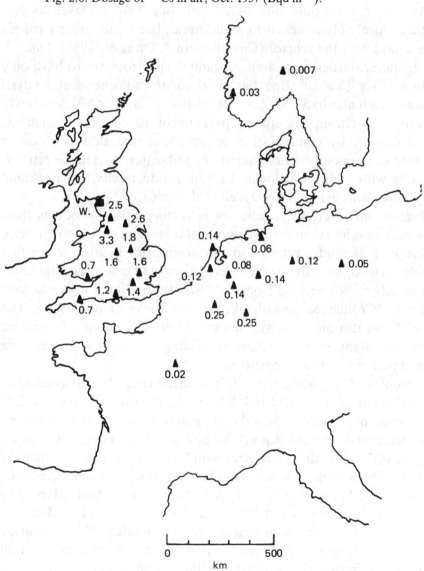

have been made. Crabtree (1959) used meteorological data (wind speed, height of inversion layer) and the measured (filter paper) activities of ^{131}I and deduced that 740 TBq (20,000 Ci) of ^{131}I crossed the line Liverpool/Flamborough Head about 140 km downwind of Windscale. To this should be added about 180 TBq (5000 Ci) deposited north of that line, giving a total of 920 TBq. Loutit *et al.* (1960) adopted the figure 740 TBq, without the allowance for deposition, for ^{131}I. Assuming a ^{131}I/^{137}Cs activity ratio of 34, they deduced the emission of ^{137}Cs as 22 TBq (600 Ci). Clarke & Macdonald (1978) used a computer prediction of release from fuel and obtained estimates of 600 TBq for ^{131}I and 46 TBq for ^{137}Cs.

The environmental filters are more likely to have been 100% efficient for ^{137}Cs than for ^{131}I. Applying Crabtree's method to the ^{137}Cs dosages leads to an estimate of 80 TBq (740 divided by 9.1) for the emission of ^{137}Cs, not allowing for the activity deposited north of the line Liverpool/Flamborough Head. The deposition velocity of ^{137}Cs was lower than that of ^{131}I, and allowing for deposition only adds about another 4 TBq.

Yet another estimate can be obtained from a measurement of ^{133}Xe in air at Wembley (Chamberlain, 1981). The activity of ^{133}Xe in the fuel which melted was 10,000 TBq (0.27 MCi) and this was easily the largest emission from Windscale, assuming as is probable that it all issued from the stack. Clarke & Macdonald (1978) showed that the measured gamma dose rate downwind during the release was consistent with the calculated dose rate from the plume, to which the dose from ^{133}Xe contributed. The activity of ^{133}Xe was measured in air liquified over the period 1400/11th to 1400/12th October (Peabody & Taylor, see Chamberlain, 1981), during which the plume passed Wembley, and was equivalent to a dosage of 220 Bqd m^{-3}. The dosage of ^{137}Cs in London air, from filter measurements, was 1.4 Bqd m^{-3} (Stewart *et al.*, 1961). Hence the ^{133}Xe/^{137}Cs ratio was 157 : 1, and the activity of ^{137}Cs in the air passing London was 10,000/157 = 64 TBq.

The activity of ^{137}Cs in the fuel which melted was 250 TBq (Chamberlain, 1981), and if the ^{137}Cs released and passing the stack filter was about 80 TBq, this represents about 30% of that in the melt zone. In the Chernobyl accident, an estimated 13% of the total inventory of ^{137}Cs in the reactor was emitted (U.S.S.R. State Committee, 1986).

The estimates of release of fission products from Windscale, which are inevitably subject to error, are not essential to the assessment of the radiological effects, which are based on local measurements of activity in milk and other foodstuffs, and in the thyroids of members of the public.

The radiological consequences were considered by Loutit *et al.* (1960) on behalf of the Medical Research Council and more recently by the National Radiological Protection Board (Crick & Linsley, 1982, 1983). Loutit *et al.* estimated the maximum doses of radiation received by the public as:

External radiation	0.6 mGy
Thyroid dose (child)	160 mGy
Thyroid dose (adult)	40 mGy

The Medical Research Council (1959) considered that 250 mGy (25 rad) was a permissible emergency thyroid dose.

It is now usual to calculate the effective dose equivalent (Appendix 1.2). The dose equivalent measured in Sieverts (Sv), takes into account the relative biological efficiency of different radiations. For gamma and beta radiation, the conversion factor is unity, but for alpha radiation it is 20. The effective dose equivalent allows also for the relative importance of irradiation of various organs to the risk of cancer. To convert thyroid dose from beta particles, measured in Gy, to effective dose equivalent, a factor 0.03 is applied. Thus the maximum thyroid doses estimated by Loutit *et al.* correspond to effective dose equivalents of 4.8 mSv (child) and 1.2 mSv (adult). Adding the external whole body gamma radiation, for which the conversion factor is unity, gives 5.4 mSv (child) and 1.8 mSv (adult).

Crick & Linsley (1982, 1983) reassessed the doses resulting from the Windscale accident, taking into account the emission of ^{210}Po (Section 1.20), which was not evaluated by Loutit *et al.* (1960). Following inhalation or ingestion, this nuclide is distributed in many tissues, and is not concentrated in a particular organ; but a factor 20 is applied to calculate the dose in Sv from its alpha particles. Crick & Linsley calculated that the collective effective dose equivalent received by people in the UK and Northern Europe, due to the Windscale accident, was 2×10^3 man-Sv. Of this total, about 40% was due to ^{210}Po. They calculated that people living in the area of maximum fallout from the accident would have received 7.1 mSv (infant), 6.8 mSv (child) or 4.1 mSv (adult), and pointed out that these doses are comparable with those received in three to four years from natural background radiation.

A cluster of cases of childhood leukemia has been found in West Cumbria (Forman *et al.*, 1987). As a result, an exhaustive re-examination has been made of the emissions from the Windscale piles and the adjacent Sellafield reprocessing plant, including those before, during and since the 1957 accident (Stather *et al.*, 1984, 1986). It was concluded that the expected number of cases of leukemia from irradiation related to the Windscale emissions was 0.016. The expected number from all radiation, including natural background and medical uses, was 0.1.

2.7 Accident to SL-1 reactor

The SL-1 (Stationary Low Power No. 1) was a 3-MW (thermal) boiling water reactor operated by military personnel at the National Reactor Testing Station, Idaho. As a result of interference with the control rods, there was an explosion on 3 January 1961 in which about 5 tonne of coolant were expelled from the pressure vessel (Horan & Gammil, 1963).

Three men were killed by the blast. Their bodies were saturated with highly contaminated water and particles of fuel had penetrated their skin. Rescue teams evacuated the casualties despite gamma dose rates up to 10 Gy h^{-1} within the reactor building.

Air samplers with filters and charcoal packs were in operation in the environs of the reactor, and samples of vegetation (sagebrush) were taken to identify the plume of deposition. It was estimated that 2.6 TBq (70 Ci) of ^{131}I were released, of which about 0.4 TBq (10 Ci) were emitted in the first 16 h after the accident and the remainder in the following 30 days (Horan & Gammil, 1963). A dosage of 1.2 Bqd m^{-3} of ^{131}I was measured at Atomic City, the nearest off-site centre of population, which was 8.5 km downwind. Essentially all the non-gaseous radioactivity was contained within the 3-acre plot round the reactor, and was estimated to include 0.02 TBq (0.5 Ci) of ^{137}Cs and 0.004 TBq (0.1 Ci) of ^{90}Sr. The release from SL-1, like those from the Windscale and Chernobyl accidents, contained predominantly volatile fission products.

2.8 Chemical explosion in USSR

On 29 September 1957, eleven days before the Windscale accident, there was a chemical explosion in a Soviet plant treating active wastes, situated in the Urals. No mention of the accident was made in Soviet media at the time, but an exiled scientist, Z. A. Medvedev collected information and published it in the west (Medvedev, 1976, 1979). A further study, using Medvedev's data and oblique references to the effects of the disaster in Soviet ecological literature, was made by scientists at Oak Ridge (Trabalka *et al.*, 1980). After the elapse of 32 a, the Soviet authorities have now published a summarised account of the effects of the accident (Nikipelov *et al.*, 1989).

The explosion took place in tanks containing highly active nitrate–acetate wastes. It resulted in the dispersion of 74 PBq (2 Megacuries) of fission product activity, about 50 times greater than the activity dispersed in the Windscale accident (excluding in the latter case ^{133}Xe and ^{3}H emitted as gases). It is not clear whether the activity dispersed was estimated from the activity lost from the tanks, or was calculated by integration of the areas within isopleths of deposition.

Fallout of activity affected large areas in the provinces of Chelyabinsk and Sverdlovsk. In the worst area, deposited activity gave an initial gamma dose of about 75 mR h^{-1} (0.6 mGy h^{-1}) which was two orders of magnitude greater than the maximum dose rate off-site at Windscale. The population in this area, numbering 600, were evacuated after 10 d, having meanwhile accumulated an average external dose of 0.17 Gy.

The activities of various fission products, and their percentage contribution to the total release, as given by Nikipelov *et al.*, are shown in Table 2.5. The small emission of ^{137}Cs, relative to ^{90}Sr, had previously been deduced by Trebalka *et al.* (1980) from references in the Soviet literature. It appears that ^{137}Cs had been separated from other fission products further upstream in the process. The relative activities of ^{90}Sr, ^{95}Zr, ^{106}Ru and ^{144}Ce in Table 2.5 are about as would be expected in fission product wastes stored for about 200 d after removal from a reactor (Trebalka *et al.*, 1980). Such a storage period would account for the absence of ^{131}I and other short-lived fission products.

The release thus consisted of the more refractory fission products,

Table 2.5. *Activities released in chemical explosion at separation plant in Urals (Nikipelov et al., 1989)*

Nuclide	Half-life	Release		Percentage of total release
		Ci	TBq	
^{89}Sr	51 d	traces		
^{90}Sr + ^{90}Y	29 a	1.1×10^5	4×10^3	5.4
^{95}Zv + ^{95}Nb	65 d	5×10^5	2×10^4	25
^{106}Ru + ^{106}Rh	1 a	7×10^4	3×10^3	3.7
^{137}Cs	30 a	7×10^2	27	0.036
^{144}Ce + ^{144}Pr	784 d	1.3×10^6	5×10^4	66

with ^{90}Sr representing 5% of the total, and was quite different in this respect from the Windscale and Chernobyl releases, in which volatile elements were predominant.

The initial hazard was gamma radiation from fallout activity, but the long-term hazard was the deposition of ^{90}Sr and its entry into food chains. Nikipelov *et al.* (1989) stated

> 'For the area with a ^{90}Sr contamination density of 0.1 Ci/km^2 (double the level of global fallout) the maximum length of the deposition track under the radioactive plume reached 300 km; for the ^{90}Sr contamination density of 2 Ci/km^2 it reached 105 km, with a width of 8–9 km.'

The term 'global fallout' appears to refer to the peak level reached in the early 1960s rather than to that in 1957.

Large-scale evacuation began after 250 days and was completed 670 days after the accident, and applied to 10,000 persons living in an area of 1000 km^2 within the 2 Ci km^{-2} (74 kBq m^{-2}) isopleth of ^{90}Sr deposition. At the limit of the evacuation zone, ingestion of ^{90}Sr had given an effective dose equivalent of 2.3 rem (23 mSv), with milk contributing 60–80% of the intake.

On the state farms, monitoring of ^{90}Sr in milk has continued to the present time. The activity has decayed with a half-life of 10 a, due to the combined effects of radioactive decay, movement down the soil profile and reduced availability of ^{90}Sr to uptake by plants. UNSCEAR (1977) deduced a half-life of 6 a from measurements of ^{90}Sr in milk in Europe and United States after the cessation of large-scale weapons testing in 1964 (Section 2.15).

Nikipelov *et al.* did not discuss the physicochemical form in which the activity was disseminated, but from the fallout pattern it appears that the particle size was small. The plume was narrow (8–9 km wide at 105 km downwind), consistent with an instantaneous explosive release. Relative to a release of 1.1×10^5 Ci (4×10^3 TBq) of ^{90}Sr, the fallout at 1.5 km and 300 km amounted respectively to 2×10^{-5} and 1×10^{-6} of the source strength per km^2. In the Windscale accident, 2.5×10^4 Ci (9×10^2 TBq) of ^{131}I were released. The deposition downwind was 0.5 Ci km^{-2} at 105 km and 0.6 Ci km^{-2} at 300 km (Chamberlain, 1959; Stewart & Crooks, 1958), these levels amounting to 2×10^{-5} and 2×10^{-6} per km^2 of the source strength. The similarity in long-range travel was no doubt fortuitous, the mechanisms of transfer to the ground being different in the two cases. The Soviet data appear consistent with most of the release being in particles in the 1–10 μm size range.

2.9 Accident at Three Mile Island

On 28 March 1979, there was loss of coolant in the core of the 2700-MW pressurised water reactor at Three Mile Island and the temperature rose to over 2000°C. The zirconium cladding of the fuel elements reacted with steam and produced an estimated 300 kg of hydrogen. Failure of the cladding released volatile fission products into the coolant. Little if any of the fuel melted. Coolant escaped into the containment building through a relief valve. Some coolant was pumped into the auxiliary building, where it escaped through a ruptured pressure disk. Large activities of isotopes of Xe and Kr escaped to atmosphere from the auxiliary building. A relatively small amount of I escaped. The auxiliary building was fitted with a charcoal filter, which was moderately effective. The main reason for the small escape of I was that the activity remained in solution in the coolant water. The containment building remained intact, despite an internal hydrogen/air explosion which raised the internal pressure to 28 psi.

Analysis of a sample of primary coolant taken on 30 March reported by English (1979) in The Kemeny Report, showed that about 10% of the inventory of rare gas, I and Cs fission products had been liberated from the fuel, but only about 0.1% of Te and less than 0.01 of alkaline and rare earths fission products (Table 2.6). Further samples of coolant taken 12 d later showed that leaching of refractory fission products had increased their concentrations by an order of magnitude. However,

Table 2.6. *Release of fission products in accident at Three Mile Island*

| Nuclide | Reactor inventory (EBq) | Fractional release from fuel | | Activity to atmosphere (TBq) |
		to coolant or containment[a]	to atmosphere	
^{133}Xe	5.6	0.4	0.016	9×10^{-4}
^{131}I	3.6	0.095	2×10^{-7}	0.6
^{137}Cs	0.03	0.11	—	—
^{132}Te	3.6	1.1×10^{-3}	—	—
^{106}Ru	0.12	4×10^{-5}	—	—
^{89}Sr	2.4	3×10^{-5}	—	—
^{140}Ba	4.4	7×10^{-5}	—	—

(*a*) As at 1 d after accident.

other investigations indicated higher releases of ^{131}I to the coolant, up to 55% of the inventory. A gas sample taken on 31 March indicated that 40% of the inventory of ^{133}Xe was in the atmosphere of the containment building.

The activities of ^{131}I and ^{129}I, in the water and gas phases of the containment building, were measured (Campbell, 1979) and gave a value about 10^5 to the partition coefficient P defined.

$$P = \frac{\text{concentration per litre liquid}}{\text{concentration per litre gas}}$$

The low volatility of the iodine was attributed to the high pH of the water (about 8), the reducing potential of the hydrogen present in the containment, and the possible presence of silver from melted control rods. The reducing conditions may also explain the limited transfer of tellurium and ruthenium to the coolant (Table 2.6), since the reduced forms of these elements are less volatile than the oxides.

2.10 Chernobyl accident

The RBMK reactor at Chernobyl had uranium dioxide fuel elements clad in zirconium, a graphite moderator and boiling-water coolant. It was rated at 3200 MW, slightly higher than the rating of the Three Mile Island reactor and 20 times higher than that of Windscale Pile No. 1. The reactor was commissioned in December 1983, and 76% of the fuel assemblies originally loaded were still in the reactor on 26 April 1986 when the accident occurred. The average burn-up was 10,000 MWd-Te and average irradiation period about 2 a.

During experimental procedures there was a power surge, the fuel overheated and the fuel assemblies were destroyed. Residual heat release raised the fuel temperatures above 2000°C. Particles of fuel were dispersed into the moderator and coolant. Chemical reactions between zirconium and steam, and between steam and graphite, generated hydrogen which combined explosively with air. A graphite fire developed, which was finally extinguished after 14 d by dropping inert material onto the exposed core from helicopters (U.S.S.R. State Committee on the Utilization of Atomic Energy, 1986). Emissions of activity continued until 6 May.

Activity was released in three ways:

   (a) Dissemination of particles of fuel shattered in the initial explo-
       sion.

  (b) Volatilisation of fission products from fuel with subsequent condensation on particles.

  (c) Combustion of graphite into which activity had been dispersed.

Table 2.7 shows the inventory of the principal fission products in the reactor, and the activities released to the atmosphere, as reported by the U.S.S.R. State Committee (1986).

At the time of the accident, the weather pattern was dominated by a ridge of high pressure centred over the north west of the USSR. The activity released in the initial explosion at 0123 (local time) on 26 April was carried north westwards towards Scandinavia (Fig. 2.7), reaching Finland late on 27 April and Sweden early on 28 April (Devell *et al.*, 1986). Emissions from Chernobyl about 1600 on 26 April followed a more westerly track (Smith & Clark, 1986), reaching southern Germany on 29 April and the UK on 2 May. This cloud again passed over the UK on 7/8 May. Emissions between 28 April and 5 May travelled south eastwards towards Kiev and the Black Sea. Some activity was carried high in the atmosphere and entered the global west–east circulation, reaching Japan on 3 May (Aoyama *et al.*, 1986) and New York on 10 May (Larsen *et al.*, 1986).

Figure 2.7 shows the dosage (time-integrated concentration) of ^{137}Cs at stations in western Europe. More recently, Gudiksen *et al.* (1989) have used world-wide measurements and calculated much higher emissions of volatile products, 89 PBq of ^{137}Cs and 1700 PBq of ^{131}I.

Figure 2.8 shows the geometric means of weekly measurements of

Table 2.7. *Activities emitted from Chernobyl (U.S.S.R. State Committee, 1986)*

Nuclide	Reactor inventory (PBq)	Release (PBq)	Percentage release	Release relative to ^{137}Cs
^{137}Cs	280	37	13	1.00
^{131}I	3200	640	20	17.3
^{132}Te	3000	450	15	12.2
^{90}Sr	200	8	4.0	0.2
^{95}Zr	4800	155	3.2	4.2
^{103}Ru	4900	140	2.9	3.8
^{140}Ba	4800	270	5.6	7.4
^{141}Ce	5700	130	2.3	3.5
$^{239+240}$Pu	2	0.06	3.0	2×10^{-3}

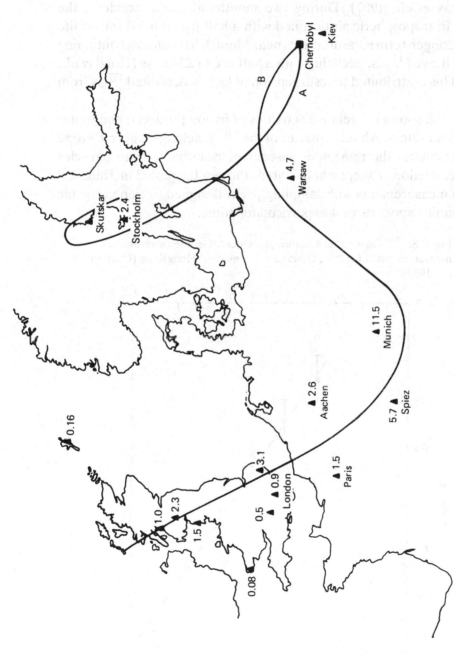

Fig. 2.7. Tracks of emissions from Chernobyl, starting at (A) 0123, and (B) 1600 local time on 26.4.86 (Smith & Clarke, 1986; Persson et al., 1987). Figures are dosages of ^{137}Cs in air (Bqd m^{-3}) (ApSimon et al., 1986; Cambray et al., 1987; Fulker, 1987; Jaworoski & Kownacka, 1988; Jost et al., 1986; Persson et al., 1987; Winkelmann et al., 1987).

^{137}Cs at Chilton (near Harwell), Gibraltar, Tromso and Hong Kong (Cambray *et al.*, 1987). During two months after the accident, the activity in tropospheric air declined with a half life of 6.3 d (mean life 9.0 d). Longer-term measurements near Munich have shown continuing air activities of ^{137}Cs, declining with a half-life of 230 days (Hötzl *et al.*, 1989). This is attributed to resuspension of locally deposited ^{137}Cs from the site.

Table 2.8 shows the relative activities of fission products, and plutonium, in air filters. About a quarter of the ^{131}I reaching western Europe was particulate, the remainder gaseous or desorbable from particles during collection. Except where stated, the results quoted in Table 2.8 refer to measurements with sampling packs designed to trap inorganic and organic vapour as well as particulate iodine.

Fig. 2.8. ^{137}Cs in air after Chernobyl accident. Geometric means of measurements at Chilton, Gibraltar, Tromso and HongKong (Cambray *et al.*, 1987).

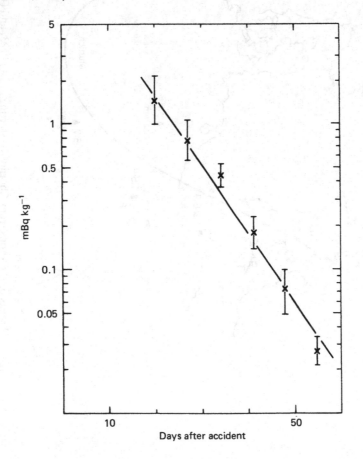

Table 2.8. *Relative activities in air filters (corrected for radioactive decay to 26.4.86)*

	Near Chernobyl[i]	Varyshevka (140 km SE) [i]	Baltic region[ii]	Munich[iii]	Harwell[iv]	New York[v]
^{137}Cs	1.00	1.00	1.00	1.0	1.00	1.00
^{131}I	10*	5*	19	14	12	7
^{132}Te	—	—	—	17	15	—
^{90}Sr	—	—	0.02	0.01	—	0.01
^{95}Zr	3.0	1.0	0.2	0.02	—	0.3
^{103}Ru	6.8	—	0.4	1.3	1.7	0.2
^{140}Ba	2.5	3.2	1.0	0.7	0.6	0.2
^{141}Ce	3.7	—	0.1	0.02	0.01	0.02
$^{239+240}$Pu	—	—	2×10^{-5}	2×10^{-5}		—

* Particulate fraction only.
References: (i) U.S.S.R. State Committee, 1986; (ii) Median of results for various locations from Krey et al. (1986), Devell et al. (1986), Cambray et al. (1987) and Aarkrog (1988); (iii) Cambray et al. (1987); (iv) Winkelmann et al. (1987); (v) Larsen et al. (1986).

There was a progressive reduction in the activities relative to Cs of the refractory fission products, isotopes of Zr and Ce (boiling points 4400 and 2900°C), with increasing distance from Chernobyl. Ru is also refractory, but its oxide RuO_4 is relatively volatile, and Ru was more persistent in long-range travel.

Most of the activity reaching western Europe was carried on particles of diameter less than 2 μm (Jost *et al.*, 1986; Winkelmann *et al.*, 1987) but larger particles reached Scandinavia (Devell *et al.*, 1986; Persson *et al.*, 1987) and these carried a higher proportion of refractory elements. Spherical particles, condensed from the vapour state, were found with activity dominated by individual elements, for example Ce, Ru, Ba or Mo (U.S.S.R. State Committee, 1986; Persson *et al.*, 1987). With longer distances of travel, the larger particles, and most of the refractory elements, were lost from the plume. Activity reaching Tennessee was found to be in particles with median diameter about 0.4 μm, similar to that of cosmogenic ^{7}Be (Bondietti & Brantley, 1986).

Fallout of activity from Chernobyl was very variable, depending on distance and bearing from the reactor, and depending very strongly on how much rain fell during passage of the plume. Figure 2.9 shows deposition of ^{137}Cs in the region of Chernobyl (Izrael, 1989). Approximately 27% of Cs and I, 60% of Zr and Ce, emitted in the accident were deposited within 80 km (U.S.S.R. State Committee, 1986). The ratio ^{90}Sr/^{137}Cs in this zone was 0.3, and a similar ratio was found by Aarkrog

Fig. 2.9. Fallout near Chernobyl: ——, 550 kBq m^{-2} (15 Ci km^{-2}) of ^{137}Cs; – – –, gamma dose rate of 42 μGy h^{-1} (5 mR h^{-1}) on 10 May 1986.

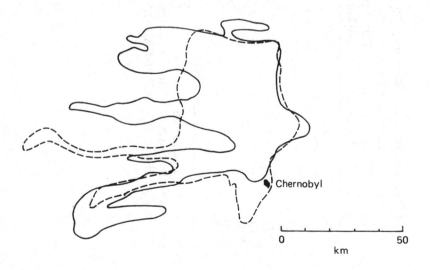

(1988) in a sample of soil from Kiev. By contrast, in fallout near Munich, $^{90}Sr/^{137}Cs$ was about 0.01 (Winkelmann *et al.*, 1987). In Russia, contamination of foodstuffs by ^{90}Sr was serious, but in western Europe fallout of ^{90}Sr was of little significance, and much less than the fallout from bomb tests in 1955–65.

Table 2.9 compares fallout at three locations and shows the effect of rainfall during passage of the plume. Fallout at Skutskar, 150 km north of Stockholm, was the highest recorded outside the USSR, deposition of ^{137}Cs being 13 times, and of ^{131}I twice, that recorded at Kiev. Aerial survey showed fallout of ^{137}Cs exceeding 100 kBq m^{-2} over an area of about 500 km^2 on the Baltic coast near Skutskar, but at Stockholm, where there was no rain, there was only about 1 kBq m^{-2} (Persson *et al.*, 1987).

In the UK, rainfall of 10–20 mm on 2–4 May gave heavy fallout in the coastal region of Cumbria and Galloway (Clark & Smith, 1988). By a coincidence, there was particularly heavy deposition in the hills 20 km SSE of Sellafield, which were also in the path of the emission of October 1957. In this area, there was about 12 kBq m^{-2} of ^{137}Cs from the Windscale accident (Fig. 2.5), 6 kBq m^{-2} from distant bomb tests, 1955–65, and 20 kBq m^{-2} from Chernobyl. Clark & Smith estimated the total fallout of ^{137}Cs over the UK at 300 TBq. This compares with about 10 TBq from the Windscale accident and 1000 TBq from weapons tests.

Table 2.9. *Effect of rainfall on fallout from Chernobyl*

	Kiev	Munich	Stutskar
Distance from Chernobyl (km)	100	1400	1300
Rainfall (mm)	Nil	6	20
Activity (kBq m^{-2})			
^{137}Cs	20	16	260
^{131}I	940	120	2000
^{95}Zr	100	0.4	18
^{140}Ba	220	14	330
Gamma dose rate 10.5.86 (μGy h^{-1})	3.2	0.32	5.0
pGy h^{-1} per Bq m^{-2} of ^{131}I	3.3	2.7	2.5

References: U.S.S.R. State Committee (1986), Winkelmann *et al.* (1987), Krey *et al.* (1986), Hardy *et al.* (1986), Webb *et al.* (1986).

Figure 2.10 shows the activity of ^{137}Cs in residents of the Harwell area from 1957 to 1988, as measured by external gamma counting *in vivo* (Kang,1989). The peak in 1987, due to fallout from Chernobyl, was much smaller than that in 1965, due to weapon tests.

Fallout at the three stations of Table 2.9 shows varying activities of I, Zr and Be relative to Cs, and this relates to the mode of deposition, as discussed later. At Kiev, and elsewhere where there was predominantly dry deposition (Roed, 1987; Cambray *et al.*, 1987), the ratio I/Cs was enhanced by transport of I vapour to ground. Sedimentation of comparatively large particles probably contributed to the fallout of refractory elements at Kiev (Aarkrog, 1988; Cambray *et al.*, 1987).

In the penultimate row of Table 2.9, estimates are given of the gamma-dose rate near the ground on 10 May 1986, 14 d after the accident. Background has been subtracted, and suitable corrections made to the dose rates recorded at Skutskar (Hardy *et al.*, 1986) and at Kiev (Webb *et al.*, 1986) where the date of measurement was not 10 May. In the last row of Table 2.9, the dose rates on 10 May have been normalised to unit deposition of ^{131}I, corrected for decay to 26 April. In Fig. 2.9, 550 kBq m^{-2} of ^{137}Cs corresponded to 42 μGy h^{-1}. The activity ratio ^{131}I/^{137}Cs, corrected for decay to 26 April, in the fallout zone within 80 km of Chernobyl was 17 (U.S.S.R. State committee, 1986). Hence the normalised gamma dose rate was 4.4 pGy h^{-1} per Bq m^{-2} of ^{131}I. In comparison with the results in Table 2.9, a higher value is to be

Fig. 2.10. ^{137}Cs in residents of Harwell district 1957–90.

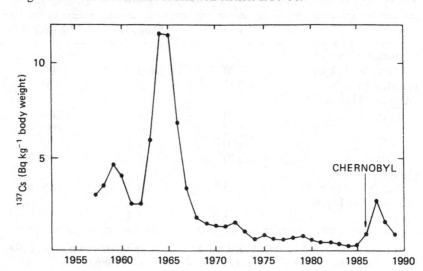

expected nearer Chernobyl, where the non-volatile fission products were more abundant, and the concordance of results gives some assurance that the recorded activities in fallout were reasonably accurate.

Table 2.10 shows a comparison of the fallout at Seascale, Cumbria, from the Windscale accident and the fallout in the Munich area from Chernobyl, together with estimates of the effective dose equivalent received by people in those areas at the relevant time (Crick & Linsley, 1982, 1983; Doerfel & Piesch, 1987). The doses at Seascale are extrapolated to 50 a from the accident to allow for long-term contributions, mainly from ^{137}Cs. The doses at Munich are for the first year only, and should be increased by about 50% to give the life-time dose (Clarke, 1987).

In comparing the two cases in Table 2.10, it is relevant to note that Seascale is 3 km from Sellafield, the site of the Windscale reactor, whereas Munich is 1400 km from Chernobyl. The calculated contributions of the various pathways were different in the two cases – namely, for adults, Windscale: ingestion 66%, inhalation 23%, external radiation 11%; Chernobyl: ingestion, 23%, inhalation 16%, external 61%.

The effective doses, per Bq of ^{131}I deposited, were higher at Seascale in 1957 than at Munich in 1986 for the following reasons:

(a) Because fallout in Munich was in heavy rain, the dosage in air was less, relative to the fallout, than at Seascale.

(b) Interception of the Chernobyl fallout by foliage was poor (Section 2.12 below) and new spring growth rapidly diluted the activity.

Table 2.10. *Fallout of activity and calculated doses*

Accident . . . Locality . . .	Windscale, 1957 Seascale, Cumbria	Chernobyl, 1986 Munich/Neuherberg F.R.G.
Fallout (kBq m^{-2})		
^{131}I	700	90
^{137}Cs	8	20
Effective dose equivalent (mSv)		
adult	4.0	0.23
child	6.4	0.26

(c) Preventive measures, and changes in eating habits of the population, reduced the intake of ^{137}Cs in the months after Chernobyl (Doerfel & Piesch, 1987).

Table 2.11 shows the calculated collective effective dose equivalents – that is the dose, multiplied by the population. For the UK, the collective doses from the Windscale and Chernobyl accidents were similar, as might be expected from a comparison of the ^{137}Cs dosages in Figs. 2.6 and 2.7. The average effective dose equivalent from Chernobyl to a UK national, extrapolated to 50 a, is about the same as the extra dose received from cosmic radiation on two flights to Spain and back! (Clarke, 1987). For the E.C. countries together, Chernobyl gave a dose about 16% of annual background, or 0.3% of the 50-a background (Table 2.11).

2.11 Deposition of fission products

The processes of deposition of particulate matter from the atmosphere will now be discussed. The mechanisms are:

(a) Sedimentation.
(b) Impaction on vegetation and other surfaces.
(c) Interception by fine roughness elements, for example hairs on leaves.
(d) Brownian diffusion to surfaces.
(e) Washout by incorporation into cloud droplets or by impaction on raindrops.

Processes (a) to (d) are classified as dry deposition, process (e) as wet deposition. Occult precipitation, which is the impaction of cloud or fog droplets onto vegetation, may be classified either way.

The relative importance of the processes depends on the particle size,

Table 2.11. *Collective effective equivalent doses over 50 a (Clarke, 1987)*

	Population	Man-Sv
Windscale accident	UK	2×10^3
Chernobyl accident	UK	3×10^3
Chernobyl accident	F.R.G.	3×10^4
Chernobyl accident	All E.C.	8×10^4
Natural background	UK	5×10^6
Natural background	All E.C.	2.5×10^7

(a) and (b) being of major importance only when the aerodynamic diameter exceeds 10 μm (Slinn, 1977). Which mechanism is dominant also depends on the distance of travel between the source of the fission product aerosol and the point of measurement. Large particles, with high sedimentation velocities, are likely to be deposited near the source. Washout becomes progressively more important with distance, as the particles diffuse into the cloud-forming layer, and this is the most effective mechanism of global fallout from weapons tests.

The washout coefficient is defined (Chamberlain, 1960)

$$W = \frac{\text{activity per kg rain}}{\text{activity per kg air}} \tag{2.1}$$

Other authors, for example Slinn (1977), have used a washout factor

$$\tilde{W} = \frac{\text{activity per unit volume rain}}{\text{activity per unit volume air}} \tag{2.2}$$

so that $W = 10^{-3} \rho_a \tilde{W}$, where ρ_a is the density of air (kg m^{-3}) at the height of measurement.

Table 2.12 shows washout ratios of radioactive and stable nuclides as measured in the UK. Fission products from distant bomb tests become attached to natural condensation nuclei in the atmosphere, and enter the accumulation mode of particle sizes (approximately 0.02 to 0.2 μm diameter). Washout ratios in the range $250 < W < 900$ ($2 \times 10^5 < \tilde{W} < 7 \times 10^5$) are usually found for these nuclei (Brenk & Vogt, 1981). Salt from sea spray is typically present as particles in the 1–5 μm range, outside the normal accumulation mode, and this explains

Table 2.12. *Washout ratios (UK measurement)*

Period	Nuclide	W	Reference
1954–60	^{137}Cs	680	Peirson & Cambray (1965)
1960–9	^{137}Cs	730	Cambray *et al.* (1970)
1961	^{140}Ba	580	Peirson & Keane (1962)
1961–4	^{210}Pb	530	Peirson *et al.* (1966)
1986	^{137}Cs	700	Clark & Smith (1988)
1972–3	Stable Pb	270	Cawse (1974)
	Br	760	
	Zn	640	
	Cl	3000	
	Na	3700	

the high values of W for Na and Cl in Table 2.12. W is also enhanced if the airborne concentration increases with height, because it is calculated with reference to the concentration near the ground whereas the washout process operates at cloud height.

In the UK, washout of ^{137}Cs from Chernobyl gave $W = 700$ ($\tilde{W} = 6 \times 10^5$) (Clark & Smith, 1988), but Persson *et al.* (1987) reported higher values of W, ranging up to 6000, for the washout of activity emitted in the early stages of the Chernobyl accident, which reached Sweden on 28 and 29 April 1986.

It is likely that both the above mentioned effects – larger particle sizes and increase in airborne activity with height – contributed to the high values of W. At later times, the airborne activity from Chernobyl was mainly submicrometre in size and had equilibrated with the accumulation mode of natural nuclei. Over the period 10–90 d from the emission, ^{137}Cs disappeared from the atmosphere with a half-life of 6 d, or mean life of 9 d (Fig. 2.8). The mass of air in the troposphere is 9000 kg per m^2 of the earth's surface, and the average daily rainfall in the northern hemisphere is 3.1 mm. Using these data, it can be deduced that the washout ratio of ^{137}Cs was

$$W = \frac{9000}{9 \times 3.1} = 330$$

Deposition other than in rain is termed dry deposition, and this includes sedimentation of particles, molecular and Brownian diffusion to surfaces, impaction on roughness elements and deposition under electrical or thermophoretic forces. The velocity of deposition is defined

$$v_g = \frac{\text{activity deposited per unit area per second}}{\text{activity per unit volume of air}} \qquad (2.3)$$

In terms of the time integral of the volumetric concentration, or dosage

$$v_g = \frac{\text{activity deposited per unit area}}{\text{dosage in air}} \qquad (2.4)$$

Strictly, v_g should be defined relative to the reference height at which the airborne concentration is measured. When this is not specified, it is understood to be about 1 m above ground. On the electrical analogy, the velocity of deposition is sometimes called the conductance, and its reciprocal the resistance.

Most measurements of fallout have been made with deposit gauges, which are similar to rain gauges. Unless arrangements are made to expose the gauge only when rain is actually falling, the deposit gauge will also collect dry deposition, that is particles deposited by any of the mechanisms listed above. Table 2.13 shows values of v_g for ^{137}Cs and ^{90}Sr calculated by dividing the annual deposit gauge collection by the average air concentration. The contribution of dry deposition to the deposit gauge is in fact usually ignored in calculating the washout ratio. Hence the velocity of deposition by washout is

$$v_g = PW\rho_a^{-1} \ \text{mm s}^{-1} \qquad\qquad (2.5)$$

where P is the rainfall rate in mm s^{-1} and ρ_a the density of air. At Harwell, from 1960–9 rainfall averaged 650 mm a^{-1} or 2.06×10^{-5} mm s^{-1}, and, with $\rho_a = 1.21$ kg m^{-3}, $W = 730$ gives $v_g = 12$ mm s^{-1} as shown in Table 2.13. Graustein & Turekian (1986) deduced v_g values for ^{90}Sr from bucket collections made at various US locations by the Environmental Measurement Laboratory, New York, and v_g values for ^{137}Cs from their own measurements on soil cores, and these are also shown in Table 2.13.

To estimate the contributions of dry deposition to the fallout of fission products, Peirson & Keane (1962), Peirson & Cambray (1965) and Cambray *et al.* (1970) exposed (a) a flat horizontal sheet of cellular acetate, area 0.30 m^2, and (b) an artificial grass surface consisting of vertical strips of cellulose acetate of height 5 cm on a horizontal sheet of area 0.23 m^2. The surfaces were exposed in the open air beneath a shelter. During periods in the autumn of 1961 and 1962 when tropospheric nuclear tests were conducted in Russia, the particle size of activity reaching the UK was larger than usual, and the dry deposition velocities were of the order 10 mm s^{-1}. From 1964–9, when the activity was derived from storage in the stratosphere of previous high-yield tests, the average dry deposition velocities of ^{137}Cs were 2 mm s^{-1} to the sheltered horizontal sheet and 3.3 mm s^{-1} to the sheltered artificial grass. Cambray *et al.* (1970) concluded that dry deposition contributed about 17% of the total deposition.

If dry deposition, or occult precipitation were a dominant mode of deposition, the activity deposited would be expected to depend on the type of vegetation cover. Graustein & Turekian (1987) deduced identical values of v_g from measurements of ^{137}Cs in soil cores and ^{90}Sr in bucket collections (Table 2.13), which indicated little effect of vegetation. At forested sites in the Appalachian Mountains, however,

Table 2.13. *Velocity of deposition of bomb-derived nuclides*

Period	Region	Nuclide	v_g (mm s^{-1})	Reference
1954–69	UK	^{137}Cs	12[a]	Cambray et al. (1970)
1954–83	US, lowland sites	^{137}Cs	14[b]	Graustein & Turekian (1986)
	mountain sites	^{137}Cs	12 to 29[b]	Graustein & Turekian (1986)
1963–9	lowland sites	^{90}Sr	14[a]	Graustein & Turekian (1986)

(a) From deposit gauge collections.
(b) From soil analyses.

impaction of cloud droplets on foliage was estimated to account for more than 30% of the deposition.

Measurements of activity in grass and soil in areas where no rain fell at the relevant time have been used to estimate dry deposition after the Chernobyl accident. In Denmark and in southern England, v_g for ^{137}Cs was about 0.5 mm s^{-1} (Roed, 1987; Clark & Smith, 1988). In Stockholm, however, where the Chernobyl fallout arrived several days earlier, and the particle size was larger, the dry deposition velocity of caesium was 5 mm s^{-1} (Persson *et al.*, 1987). Refractory elements such as ^{95}Zr had dry deposition velocities about 20 mm s^{-1}.

2.12 Interception by foliage

Whether fallout is by washout or dry deposition, a certain fraction of the deposited activity is intercepted by foliage. Foliar deposition, followed by uptake by grazing animals, is an important pathway into food chains. The leaf area of herbage eaten by cattle and sheep is large. Also, animals eat herbage as it is, whereas humans usually wash leaf vegetables and discard outer leaves, pods and husks.

Herbage can be considered as a coarse filter interposed between atmosphere and ground. If a proportion p of the activity in fallout is intercepted by foliage, it is reasonable to write

$$1 - p = \exp(-\mu M_H) \tag{2.6}$$

where M_H is the density of herbage (kg dry weight per m^2 of ground) and μ is an interception coefficient with units kg m^{-2} (Chamberlain, 1970). When μM_H is less than about 0.4 equation (2.6) can be simplified to

$$\mu = p/M_H \tag{2.7}$$

It might be expected that μ would depend on the chemical and physical form of the activity, the particle size, meteorological conditions, and type of vegetation. However, experimental field data on interception by grasses, cereals, and other narrow leaved foliage have given reasonably consistent results with a variety of radioactive and stable tracers (Table 2.14). There does not seem any consistent difference whether the activity is applied as a dry particulate or in fine droplets, provided that the volume of water is not so large as to give immediate run-off. The results for broad-leaved plants and shrubs are more scattered (Miller, 1980).

Table 2.14. *Interception coefficient of grasses and cereals*

Crop	Activity	Mode of application	μ (m^2 kg^{-1}) mean	S.E.	Reference
Mixed grasses	^{89}Sr	Spray (0.171 m^{-2})	3.33	0.56	Milbourn & Taylor (1965)
Mixed grasses	^{85}Sr	Spray (1.21 m^{-2})	2.30	0.08	Chadwick & Chamberlain (1970)
Mixed grasses	^{131}I	Vapour	2.78	0.14	Chamberlain (1970)
Mixed grasses	^{131}I	Attached to *Lycopodium* spores (30 μm)	3.08	0.15	Chamberlain (1970)
Corn	^{238}Pu	Particulate fallout from stack	3.60	0.05	Pinder et al. (1987)
Mixed grasses	^{134}Cs	Quartz particles 44–88 μm diam. applied from spreader	2.62	0.08	Peters & Witherspoon (1972)
Fescue, Lespedeza	^{95m}Tc	Spray	1.06	0.08	Hoffman et al. (1982)

Lower values of μ have been found when the particle size is greater than about 50 μm, because large particles tend to bounce off leaf surfaces, and are retained only in sites such as the axil, the angular region where the leaf joins the stalk. In pastures, most such sites are in basal tissues below the grazing level. Russell & Possingham (1961) at Maralinga, and Romney et al. (1963) in Nevada exposed herbage to fallout from ground-level nuclear explosions. The particle size distribution was wide, depending on the distance from the explosion, with most of the activity in particles greater than 50 μm, and interception was poor, corresponding to μ in the range 0.04–0.6 $m^2 kg^{-1}$.

The interception by foliage of activity in rain has been studied experimentally by Hoffman et al. (1989). Polystyrene microspheres, of diameter 3, 9 and 25 μm were labelled with ^{141}Ce, ^{95}Nb and ^{85}Sr respectively. Artificial rain, with the particles in suspension, was applied to plots of clover, fescue and mixed herbage, and the fractional interception of the activity by the foliage was measured.

Table 2.15 shows some of the results, analysed using (2.6) to derive values of μ. The results refer to two amounts of rainfall, 1 and 10 mm, applied at intensities between 14 and 32 mm h^{-1}. Only minor differences were seen in the results for different crops. The values of μ for the three sizes of particles were not significantly different. They were approximately halved when the amount of rainfall was increased tenfold.

The greatest amounts of fallout from Chernobyl were found in those areas where heavy rain fell as the plume arrived, and the fractional interception by foliage appears to have been low in these circumstances. Near Aberdeen the fallout of ^{137}Cs was 2.7 kBq m^{-2}, mostly concentrated in a heavy rainstorm on 3 May 1986 (Martin et al., 1988). Activity of ^{137}Cs in grass, extrapolated back to the time of deposition, was 800

Table 2.15. *Interception by foliage of particulate activity deposited in rain* (analysis of results of Hoffman et al., 1989)

	No. of exps.	20-μm particles	9-μm particles	3-μm particles
		μ ($m^2 kg^{-1}$), mean $\pm$ S.E.		
1 mm rainfall	7	4.9 $\pm$ 1.3	4.5 $\pm$ 0.7	4.0 $\pm$ 0.6
10 mm rainfall	12	2.8 $\pm$ 0.2	2.3 $\pm$ 0.2	2.1 $\pm$ 0.2

Bq per kg dry weight, giving $\mu = 0.3$ m^2 kg^{-1}. Results from Munich (Winkelmann et al., 1987) and from Cumbria (Fulker, 1987) were given in terms of fresh weight of grass. Assuming a fresh/dry weight ratio of 4, they correspond to μ in the range 0.4–0.8 m^2 kg^{-1}. The fraction of rainwater intercepted by foliage is much less if the rainfall is intense, and the same evidently applies to activity deposited in rain.

2.13 Field loss of activity from foliage

A number of processes, including leaching by rain, abrasion between leaves, and shedding of cuticular wax, remove activity from foliage. During the growing season new growth and die-back of senescent leaves reduce the activity per unit mass. The emergence of new spring growth reduced the activity per kg in foliage after the Chernobyl accident.

Separate rate constants of field loss λ_G and λ_H are defined according as the activity is expressed per unit area of ground or per unit biomass of herbage. Table 2.16 shows values of λ_G and λ_H obtained from measurements of activity in herbage after application in low-volume aqueous sprays. Most investigators have found little difference in field loss rates depending on the nuclide used (Milbourn & Taylor, 1965; Russell, 1965; Aarkrog & Lippert, 1971). It does not appear that leaching by rain is a major factor. Nor does volatility appear critical, since a number of determinations of λ_G for ^{131}I have given results about 0.05 d^{-1}, similar to the value found with non-volatile nuclides (Section 3.9). An important factor in field loss, whether measured as λ_G or λ_H, is the rate of growth of the vegetation (Chamberlain, 1970; Aarkrog & Lippert, 1971). The rate of loss is less when the foliage is dormant. This is consistent with the observation of Moorby & Squire (1963) that activity is lost by the shedding of wax from growing leaves.

Field-loss rates have also been measured after dry deposition of particulate activity. Witherspoon & Taylor (1970, 1971) and Peters & Witherspoon (1972) used silicaceous particles, labelled with ^{134}Cs, in three size ranges, 1–44 μm, 44–88 μm and 88–175 μm. Millard et al. (1983) used submicrometre particles labelled with ^{141}Ce or ^{134}Cs. All reported field loss as showing an initial rapid phase lasting a few days, followed by one or more slow phases. In the period from a few days to about five weeks after application of the activity, the values of λ_G were in the range 0.02–0.06 d^{-1}, with no obvious correlation with particle size, or with the incidence of rainfall.

Table 2.16. *Field loss of nuclides from herbage and cereals*

Crop	Nuclide	No. of exps.	Period of exps. (d)	Rate coefficient (d^{-1})				Reference
				per unit area λ_G		per unit mass λ_H		
				mean	S.E.	mean	S.E.	
Mixed grasses	^{89}Sr	8	28	0.054	0.005	0.080	0.005	Milbourn & Taylor (1965)
Mixed grasses summer	^{85}Sr	2	63	0.037				Chadwick & Chamberlain
winter		2	63	0.014				Chadwick & Chamberlain (1970)
Barley, rye	^{85}Sr	4	68	0.037	0.002			Aarkrog (1969)
Barley, rye	^{141}Ce	4	68	0.023	0.002			Aarkrog (1969)
Fescue, Lespedeza	^{95m}Te	4	94	0.028	0.003	0.041	0.002	Hoffman *et al.* (1982)

2.14 Normalised specific activity

If fallout of activity continues daily at a constant rate, the transfer factor from fallout to herbage, termed the Normalised Specific Activity by Chamberlain (1970), is

$$\text{NSA} = \frac{\text{activity per kg dry weight of crop}}{\text{activity deposited per m}^2 \text{ of ground per day}} \tag{2.8}$$

If the simple relation between p and μ (equation (2.7)) is assumed to hold, and the effect of dilution by growth of foliage is neglected

$$\text{NSA} = \mu/\lambda_G \qquad \text{m}^2\text{d kg}^{-1} \tag{2.9}$$

If, more realistically, the biomass of the crop is assumed to increase linearly with time for T_0 days, and fallout is constant over that period, then it is easily shown (Chamberlain, 1970) that

$$\text{NSA} = \frac{\mu}{\lambda_G}\left[1 - \frac{1 - \exp(-\lambda_G T_0)}{\lambda_G T_0}\right] \tag{2.10}$$

Table 2.17 shows values of NSA from equation (2.10) with $\mu = 3 \text{ m}^2 \text{ kg}^{-1}$, $\lambda_G = 0.05 \text{ d}^{-1}$ or 0.03 d^{-1} and $T_0 = 30, 50$, or 80 days.

Chamberlain (1970) derived values of the NSA by comparing the activity of ^{90}Sr in herbage with the monthly fallout during 1955–8. At that early stage in the series of large bomb tests, the cumulative fallout was much less than it became subsequently, and the contribution of uptake from soil to ^{90}Sr in herbage was small. Table 2.18 shows NSA values from Berkshire (Harwell survey) and from other areas (Letcombe Laboratory survey), as given by Chamberlain (1970). NSA values were higher in autumn and winter than in spring and summer, and higher in upland than lowland areas, the higher values corresponding to lower rates of growth of herbage. Also, upland herbage often has needle-like leaves which act as efficient filters for deposited material (Osburn, 1966).

Table 2.17. *Normalised specific activity of herbage* ($\mu = 3 \text{ m}^2 \text{ kg}^{-1}$)

Period of growth, T_0, (d) ...	30	50	80
NSA (m²d kg⁻¹)			
$\lambda_G = 0.05 \text{ d}^{-1}$	29	38	45
$\lambda_G = 0.03 \text{ d}^{-1}$	34	48	62

NSA values for ^{90}Sr are similar to those for ^{137}Cs. Measurements from Slovakia (Csupka *et al.*, 1966) give values in the range 20–25 m^{2}d kg^{-1} for both nuclides.

Pinder *et al.* (1985) deduced NSA values in the range 20–41 m^{2}d kg^{-1} from measurements of ^{238}Pu in corn near the Savannah River Plant. Similar values also apply to non-radioactive trace elements. Allcroft *et al.* (1965) made a series of measurements of fluorine in herbage on a farm near Stoke-on-Trent and also recorded fallout of fluorine in a nearby deposit gauge. During the period of study (1956–61) the productivity of the pasture was improved, and the yearly average NSA declined from 59 to 27 m^{2}d kg^{-1} (Chamberlain, 1970). When productivity is low, the time T_0 before maturity is increased, and so is the NSA (Table 2.17). Also λ_G may be increased, since the rate of field loss seems to be related to growth rate.

Lichens and bryophytes grow very slowly, and are adapted to absorb nutrients from the atmosphere. NSA values of order 1000 m^{2}d kg^{-1} can be deduced from data of Andersen *et al.* (1978) on the accumulation of metals by bryophytes in the Copenhagen district. In the Arctic, lichens form the basis of a food chain via reindeer to man, and the concentration of ^{137}Cs in the lichens has given exceptionally high body burdens of ^{137}Cs in Eskimos and Laplanders (UNSCEAR, 1967).

The parts of crops which are used for direct human consumption are mostly protected from foliar contamination by husks, pods or outer leaves, and these are discarded during preparation of food. Some activity is transferred from external to internal plant tissues, more so for Cs than for Sr (Middleton, 1959). Simmonds & Linsley (1982), analys-

Table 2.18. *NSA of* ^{90}*Sr in herbage*

Location	Season	NSA (m^{2}d kg^{-1})	
		mean	S.E.
Berks	Summer 1957	32	4
	Winter 1957/8	84	11
Lowland sites	May/July 1958	49	4
	Aug/Oct 1958	97	15
Upland site	July 1958	116	
	Aug. 1958	167	
	Oct. 1958	179	

ing data from UK, USA and Denmark, found NSA of grain in the ranges 2–9 (Sr) and 5–12 (Cs) $m^2 d\,kg^{-1}$.

2.15 Transfer to milk

If there is fallout from nuclear weapons tests, or from reactor accidents, milk is likely to be the predominant source of ^{90}Sr in diet, and an important source of ^{137}Cs. A few days will intervene between an episode of fallout and the peak concentration in milk, giving time for preventative measures. It is therefore important to know the transfer factors, so that the future levels in milk can be predicted from the amount of fallout. The transfer depends on the retention of the fallout on the foliage, the amount of herbage eaten by the cattle, and the fraction of the cow's daily intake secreted per litre of milk. The transfer factor feed/milk is defined

$$F_m = \frac{\text{activity per litre of milk}}{\text{activity ingested per day}} \tag{2.11}$$

Garner (1971) and Ng (1982) have reviewed laboratory and field data on F_m. For strontium isotopes, it depends on the calcium content of the feed and the milk, and some time may elapse before an increased intake of Sr is fully reflected in milk. Sr, like Ca, is stored in the skeleton, and 50% or even more of Sr in milk may derive from bone, depending on the cow's calcium status (Comar et al., 1961).

There is a seasonal effect in transfer of Cs to milk, with F_m increasing in autumn (Van den Hoek et al., 1969). Transfer to milk may be greater when Cs is absorbed from soil as compared with foliar contamination (Ward et al., 1989). After the Chernobyl accident F_m for ^{137}Cs in West Cumbria was $4 \times 10^{-3}\,d\,l^{-1}$ in early May, increasing to $9 \times 10^{-3}\,d\,l^{-1}$ by mid-June (Jackson et al., 1987).

In the sequel, F_m will be taken as $1 \times 10^{-3}\,d\,l^{-1}$ for ^{90}Sr and $6 \times 10^{-3}\,d\,l^{-1}$ for ^{137}Cs, these being about the medians of reported values (Ng, 1982). Variations by a factor of two or three either way are to be anticipated. When the Sr/Ca ratio in milk is compared with the same ratio in the feed of cattle, a discrimination against Sr by a factor 10 is observed (Comar, 1966a), but there is little discrimination against Cs relative to K (Booker, 1959; Sansom, 1966) and this difference is reflected in the relative values of F_m for Sr and Cs.

The concept of the transfer factor can be taken back a stage by defining

$$k_m = \frac{\text{activity per litre milk}}{\text{activity deposited per m}^2 \text{ per day}} \qquad (2.12)$$

The units of k_m are $\text{m}^2\text{d l}^{-1}$.

If cattle consume G kg dry weight of herbage per day

$$k_m = \text{NSA} \times F_m \times G \qquad (2.13)$$

G depends on the type of herbage and on the way the herd is managed. For UK conditions, Garner (1960) suggested 9 kg d^{-1}, and for US conditions Comar (1966b) recommended 15 kg d^{-1}. Assuming NSA $= 30$ $\text{m}^2\text{d kg}^{-1}$, typical of good grazing in summer (Tables 2.17, 2.18), $G = 12$ kg d^{-1}, and the values of F_m given above, k_m is 0.36 $\text{m}^2\text{d l}^{-1}$ for ^{90}Sr and 2.2 $\text{m}^2\text{d l}^{-1}$ for ^{137}Cs (Table 2.19).

The transfer factor can also be calculated as the integral over time of the activity in milk after a single episode of fallout. Van den Hoek *et al.* (1969) applied aqueous solutions of ^{85}Sr and ^{134}Cs as low-volume spray to a pasture at Mol, Belgium. Cattle were fed on the grass under three different regimes, free grazing, rotational grazing, and cut foliage. Figure 2.11 shows the average concentrations in milk normalised to unit deposition of activity. Also shown are curves calculated by Linsley *et al.* (1986) for hazard assessment. The normalised activity of ^{137}Cs in milk at Seascale, Cumbria after the Windscale accident was consistent with the curve in Fig. 2.11 (Chamberlain, 1987) but a lower transfer factor, 0.05 m^2l^{-1} at the maximum, applied to the fallout from Chernobyl (Fulker, 1987). In the Karlsruhe district of Germany, the mean concentrations of ^{137}Cs in milk in May and June 1986 were 8.8 Bq l^{-1} and 10.5 Bq l^{-1}, and the fallout was 1.7 kBq m^{-2}, giving transfer factors of 5 and 6×10^{-3} m^2l^{-1}. It was noted in Section 2.12 that the Chernobyl fallout was poorly intercepted by foliage.

Table 2.19. *Transfer coefficients from fallout to milk following foliar uptake*

Derivation	k_m (m^2 d l^{-1})	
	^{90}Sr	^{137}Cs
NSA $= 30$ m^2d kg^{-1}; $G = 12$ kg d^{-1},		
$F_m = 1 \times 10^{-3}$ d l^{-1} (^{90}Sr)	0.36	
$F_m = 6 \times 10^{-3}$ d l^{-1} (^{137}Cs)		2.2
From areas under curves of Fig. 2.11	0.27	2.2

The slope of the curves in Fig. 2.11, from day 10 onwards, corresponds to a field-loss rate of $0.058 \, \mathrm{d}^{-1}$ (compare Table 2.16). Assuming this continues indefinitely, the integrals under the curves are 0.27 $\mathrm{m^2 d \, l^{-1}}$ for $^{90}\mathrm{Sr}$ and 2.2 $\mathrm{m^2 d \, l^{-1}}$ for $^{137}\mathrm{Cs}$, values of k_m which are compared in Table 2.19 with those calculated previously.

Transfer coefficients can also be deduced from statistics of the levels of $^{90}\mathrm{Sr}$ and $^{137}\mathrm{Cs}$ in milk measured during monitoring of the fallout from nuclear weapon tests. This has the advantage over experimental work of taking into account variations in agricultural practices, but introduces complications such as the use of root crops and imported feed for cattle and the uptake by crops from the soil of activity deposited in previous years. UNSCEAR (1977) analysed data from a number of countries, obtained in the years 1958–74, in the form,

$$C(i) = b_1 A(i) + b_2 A(i-1) + b_3 \sum_{m=1}^{\infty} \exp(-\mu m) A(i-m)$$

(2.14)

where $C(i)$ = activity in milk in the ith year, $A(i)$ = fallout per unit area in the ith year, b_1, b_2, b_3 = empirical constants, and μ = decay factor.

Fig. 2.11. Transfer of Cs and Sr from fallout to milk. Experimental results of Van den Hoek et al. (1960). Lines are as calculated by Linsley et al. (1986).

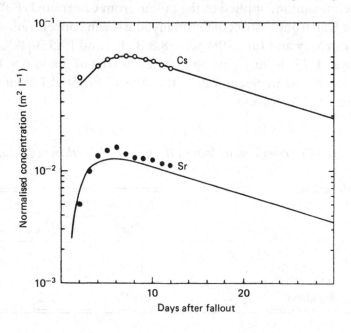

The first term on the right-hand side of equation (2.14), termed the rate factor, represents the direct contamination of herbage by fallout during the growing season. The second term, lag rate factor, is the contribution from the previous year's fallout. This includes the contribution of uptake from the surface soil and matt and also the effects of carry-over of silage and other feeding stuffs from one year to the next. The third term, soil factor, represents the contribution of root uptake, allowing for radioactive decay and reduced availability as nuclides move down the soil profile and become fixed to clay minerals.

In the UNSCEAR report the activities of ^{90}Sr and ^{137}Cs in milk are given per gram calcium and per gram potassium respectively, and the coefficients in equation (2.14) have the units 10^{-3} km^2 a (g Ca)$^{-1}$ and 10^{-3} km^2 a (g K)$^{-1}$. Assuming that milk contains 1.2 g Ca and 1.6 g K per litre, the UNSCEAR units are converted to m^{2}d l^{-1} by multiplying by 0.44 for ^{90}Sr and 0.58 for ^{137}Cs, and some of the results are shown in Tables 2.20 and 2.21.

The b_1 coefficients are analogous to k_m, and are higher for ^{137}Cs than for ^{90}Sr because F_m is higher for ^{137}Cs. The data for the Faroes in Tables 2.20 and 2.21 show the increased transfer factor where the climate is rather adverse. The lower productivity and slower growth of herbage account for the difference (Aarkrog, 1971). UNSCEAR (1977) also give a time-integrated transfer factor (P_{23} in their notation) which is essentially the sum to infinity of $C(i)$ in equation (2.14) assuming $A(i)$ is

Table 2.20. *Parameters of transfer function of ^{90}Sr from fallout to milk (m^{2}d l^{-1}) (UNSCEAR, 1977)*

	UK	US	Denmark	Faroes
b_1	0.39	0.38	0.43	1.19
b_2	0.21	0.10	0.20	0.56
B	1.07	1.10	1.37	3.28

Table 2.21. *Parameters of transfer function of ^{137}Cs from fallout to milk (m^{2}d l^{-1})*

	UK	US	Denmark	Faroes
b_1	1.02	1.33	1.38	3.97
b_2	0.94	0.78	0.40	3.18
B	2.77	2.12	1.89	16.07

constant. This sum, denoted B, with units $m^2 d \, l^{-1}$ is shown in Tables 2.20 and 2.21. The excess of B over $b_1 + b_2$ shows the contribution of root uptake compared with direct contamination of foliage, which is further considered in the next section.

2.16 Transfer of fission products from soil to crops

The factors which determine uptake from soil are complex. Among the variables are:

   (a) The profile of activity with depth in the soil, as modified by leaching and by cultivation.

   (b) Effect of soil parameters, including pH, cation exchange capacity, and percentage organic matter in soil.

   (c) Types of crop, methods of cultivation and harvesting.

Uptake from soil is often expressed as a concentration factor (activity per kg crop divided by activity per kg soil). This is appropriate if the activity is well mixed in the soil, but activity from fallout, particularly on permanent pastures, is not well mixed, and it is convenient to calculate the activity in the crop normalised to unit deposition. This gives the same units ($m^2 \, kg^{-1}$) as the interception coefficient (Table 2.14).

During the 1960s, much attention was paid to the fallout of radio-strontium from bomb tests. Ellis *et al.* (1968) applied ^{89}Sr in solution as a low-volume spray to plots of rye-grass at seven locations on varying types of soil. The activity was applied in spring, and the rye-grass was harvested in three cuts during the year. The activity in the first cut was mainly the remaining part of the original foliar contamination. That in the remaining cuts derived by uptake from the soil or from the surface matt. Figure 2.12 shows activity per kg dry weight in the second and third cuts of grass, normalised to unit deposition, and plotted against the exchangeable Ca in the soil. The effect of soil Ca on the uptake is evident, and would be more pronounced if the activity had been expressed per g Ca instead of per kg dry weight in the foliage, since the Ca per kg dry weight in foliage was correlated with the Ca in soil. Ellis *et al.* also studied the effects of ploughing and resowing after application of the [89]Sr. This treatment reduced the uptake by a factor of about three. In other experiments, Squire (1966) showed that successive harvests of rye-grass gave declining uptake of ^{90}Sr, even without cultivations. The decline amounted to a factor two or three over seven years, and was attributed to downward movement of activity in the soil.

The range of values of uptake of Sr from soil in Fig. 2.12 is 10^{-2} to $10^{-1} \, m^2 \, kg^{-1}$, and this compares with the range 1 to 4 $m^2 \, kg^{-1}$ of the interception coefficient in Table 2.14, showing that the initial foliar contamination is normally much greater than the subsequent uptake from soil. In terms of the time-integrated concentration in foliage, uptake from soil is predominant. The time-integrated transfer factors B in Table 2.20 are about twice the sum of the foliar contamination factors b_1 and b_2.

Uptake of ^{137}Cs from soil was also studied in the 1960s (Frederiksson *et al.*, 1966) and there has been renewed interest in relation to the long-term effects of the Chernobyl accident. The uptake is very dependent on the type of soil. In normal agricultural soils, Cs, like K, is fixed on the cation-exchange sites in the clay fraction. Within plants, Cs is more mobile than Sr, and it is translocated more readily from leaves to seeds (Middleton, 1959). It can also be absorbed from the matt horizon at the top of the soil, as this comprises organic material, but uptake from normal soils becomes small once Cs is fixed on the clay fraction. In Table 2.21, the data for UK, US and Denmark shows that B only slightly exceeds $b_1 + b_2$.

The situation is different when the clay fraction is very small, as is the

Fig. 2.12. Uptake of ^{89}Sr by rye-grass from soil (Ellis *et al.*, 1968).

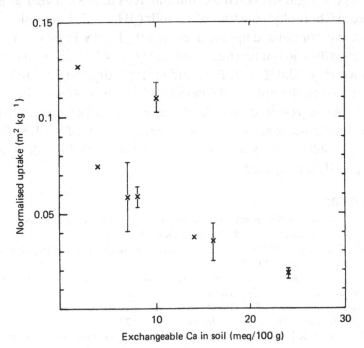

case in highly organic and also in lateritic soils (Frederiksson *et al.*, 1966). Barber (1964) grew rye-grass in pots with soil uniformly contaminated with ^{137}Cs, and found concentration factors ranging from 0.08 to 2.9, which were correlated with the cation-exchange capacity of the organic fraction of the soil. In further experiments, Barber & Mitchell (1964) applied ^{137}Cs and ^{90}Sr in winter to turves which had been transplanted from various types of pasture. The normalised uptake of ^{137}Cs ranged from 0.02 to 0.09 m^2 kg^{-1} dry weight in the first and from 0.003 to 0.06 m^2 kg^{-1} in the second subsequent season. There was strong correlation with the organic exchange capacity, and, on the most peaty soil, the uptake of ^{137}Cs in the second year exceeded that of ^{90}Sr. The soils in the Faroes are peaty, and the effect can be seen in the high value of B in Table 2.21, but, in most western countries, peaty soils are confined to upland districts, which carry beef cattle and sheep, but are not used for production of milk.

The heaviest fallout in the UK from the Chernobyl accident was in upland districts, and occurred in heavy rain. The initial interception by foliage was comparatively slight (μ about 0.3 to 0.8 m^2 kg^{-1} dry weight: Section 2.12 above), so the activities in herbage in 1986, though high relative to other regions of the UK, were not as high as might have been expected from the fallout. Sandalls *et al.* (1989) measured ^{134}Cs in soils and herbage at eight sites on the Cumbrian fells in 1987, a year after the fallout from Chernobyl. On four peaty soils (pH 4.3–5.4, organic matter 83–88%), the normalised uptake averaged 0.11 ± (S.E.) 0.02 m^2 kg^{-1}. On four equally acid but less peaty soils (pH 4.2–4.7, organic matter 19–38%), it averaged 0.038 ± (S.E.) 0.015 m^2 kg^{-1} dry weight. On normal agricultural soils, the uptake of Chernobyl ^{137}Cs in 1987 was about 10^{-3} m^2 kg^{-1}. As a result of uptake from peaty soils, the activity of radiocaesium in herbage and in sheep from hill areas of the UK did not decline as rapidly in the years following the Chernobyl accident as had been originally anticipated.

References

Aarkrog, A. (1969) On the direct contamination of rye, barley, wheat and oats with ^{85}Sr, ^{134}Cs, ^{54}Mn and ^{141}Ce. *Radiation Botany*, **9**, 357–66.
  (1971) Prediction models for strontium-90 and caesium-137 levels in the human food chain. *Health Physics*, **20**, 297–311.
  (1988) The radiological impact of the Chernobyl debris compared with that from nuclear weapons fallout. *Journal of Environmental Radioactivity*, **6**, 151–62.
Aarkrog, A. & Lippert, J. (1971) Direct contamination of barley with ^{51}Cr, ^{59}Fe, ^{58}Co, ^{65}Zn, ^{203}Hg and ^{210}Pb. *Radiation Botany*, **11**, 463–72.
Allcroft, R., Burns, K.N. & Herbert, C.H. (1965) Fluorosis in cattle – development and

alleviation, experimental studies. Ministry of agriculture, Fisheries and Food, Animal Disease Survey Report 2, Part 2, HMSO, London.

Andersen, A., Hormand, M.F. & Johnson, I. (1978) Atmospheric heavy metal deposition in the Copenhagen area. *Environmental Pollution*, **17**, 133–51.

Aoyama, M., Hirose, K., Suzuki, Y., Inoue, H. & Sugimura, Y. (1986) High level radioactive nuclides in Japan in May. *Nature*, **321**, 819–20.

ApSimon, H.M., Macdonald, H.F. & Wilson, J.J.N. (1986) An initial assessment of the Chernobyl-4 reactor accident release source. *Journal of Society for Radiological Protection*, **6**, 109–19.

Barber, D.A. (1964) Influence of soil organic matter as the uptake of caesium-137 by perennial rye grass. *Nature*, **204**, 1326.

Barber, D.A. & Mitchell, W.A. (1964) The entry of caesium-137 and strontium-90 into the herbage of permanent pasture. In: ARCRL Report 12. Agricultural Research Council Radiological Laboratory. Letcombe, Oxon., pp. 59–61.

Bondietti, E.A. & Brantley, J.N. (1986) Characteristics of Chernobyl radioactivity in Tennessee. *Nature*, **322**, 313–14.

Booker, D.V. (1959) Caesium-137 in dried milk. *Nature*, **183**, 921–4.
   (1962) Caesium-137 in soil in the Windscale area. AERE Report R-4020, Harwell, Oxon.

Brenk, H.D. & Vogt, K.J. (1981) The calculation of wet deposition from radioactive plumes. *Nuclear Safety*, **22**, 362–71.

Cambray, R.S., Fisher, E.M.R., Brooks, W.L. & Peirson, D.H. (1970) Radioactive fallout in air and rain: results to the middle of 1970. AERE Report R-6556, HMSO, London.

Cambray, R.S. *et al.* (1987) Observations on radioactivity from the Chernobyl accident. *Nuclear Energy*, **26**, 77–101.

Campbell, D.O. (1979) Behaviour of iodine under accident conditions at Three Mile Island. *The Accident at Three Mile Island* (*Kemeny Report*). Reports of the technical assessment task force, Vol II, Appendix F. U.S. Govt. Printing Office, Washington D.C.

Cawse, P.A. (1974) A survey of atmospheric trace elements in the UK (1972–73). AERE Report R-7669, HMSO, London.

Chadwick, R.C. & Chamberlain, A.C. (1970) Field loss of radionuclides from grass. *Atmospheric Environment*, **4**, 51–6.

Chamberlain, A.C. (1959). Deposition of ^{131}I in northern England in October 1957. *Quarterly Journal of the Royal Meteorological Society*, **85**, 350–61.
   (1960) Aspects of the deposition of radioactive and other gases and particles. *International Journal of Air Pollution*, **3**, 63–88.
   (1970) Interception and retention of radioactive aerosols by vegetation. *Atmospheric Environment*, **4**, 57–78.
   (1981) Emission of fission products and other activities during the accident to Windscale Pile No. 1 in October 1957. AERE Report M-3194. Harwell, Oxon.
   (1987) Environmental impact of particles emitted from Windscale piles, 1954–7. *The Science of the Total Environment*, **63**, 139–60.

Chamberlain, A.C. & Dunster, H.J. (1958) Deposition of activity in northwest England from the accident at Windscale. *Nature*, **182**, 629–30.

Clark, M.J. & Smith, F.B. (1988) Wet and dry deposition of Chernobyl releases. *Nature*, **332**, 245–9.

Clarke, R.H. & Macdonald, H.F. (1978) Radioactive releases from nuclear installations. *Progress in Nuclear Energy*, **2**, 77–152.

Clarke, R.H. (1987) Dose distribution in western Europe following Chernobyl. In: *Radiation and Health*, ed. R. Russell Jones & R. Southwood. John Wiley, Chichester, pp. 251–62.

Comar, C.L. (1966a) Transfer of strontium-90 into animal produce. In: *Radioactivity and Human Diet*, ed. R. Scott Russell. Pergamon, Oxford, pp. 247–319.

(1966b) Radioactive materials in animals, entry and metabolism. *Ibid.*, pp. 127–57.

Comar, C.L., Wasserman, R.H. & Twardock, A.R. (1961) Secretion of calcium and strontium into milk. *Health Physics*, **7**, 69–80.

Crabtree, J. (1959) The travel and diffusion of the radioactive material emitted during the Windscale accident. *Quarterly Journal Royal Meteorological Society*, **85**, 362–70.

Crick, M.J. & Linsley, G.S. (1982), (1983). An assessment of the radiological impact of the Windscale reactor fire, October 1957. NRPB report R-135, and Addendum (1983). HMSO, London.

Crooks, R.N., Glover, K.M., Haynes, J. W., Osmond, R.G. & Rogers, F.J.G. (1959) Alpha activity on air filter samples collected after the Windscale accident. AERE Report R-2952. HMSO, London.

Crouch, E.A.C. (1977) Fission product yields from neutron-induced fission. *Atomic Data & Nuclear Data Tables*, **19**, 417–532.

Crouch, E.A.C. & Swainbank, I.G. (1958) Radiochemical and physical examination of debris from the Windscale accident. AERE Report C/R 2589. Harwell, Oxon.

Csupka, S., Petrasova, M. & Carach, J. (1966) Seasonal variation in the concentration of caesium-137 in grass and alfalfa. *Nature*, **213**, 1204–6.

Devell, L., Tovedal, H., Bergström, V., Appelgren, A., Chyssler, J. & Andersson, L. (1986) Initial observations of fallout from the reactor accident at Chernobyl. *Nature*, **321**, 192–3.

Devell, L., Aakrog, A., Blomqvist, L., Magnussen, S. & Tveten, V. (1986) How the fallout from Chernobyl was detected and measured in the Nordic countries. *Northern Europe*, **11**, 16–17.

Drevinsky, P.J. & Pecci, J. (1965) Size and vertical distributions of stratospheric radioactive aerosols. In: *Radioactive Fallout from Nuclear Weapons Tests*, ed. A.W. Klement Jr. CONF 765. U.S. Dept of Commerce. N.T.I.S. Springfield, Va.

Doerfel, H. & Piesch, E. (1987) Radiological consequences in the Federal Republic of Germany of the Chernobyl reactor accident. *Radiation Protection Dosimetry*, **19**, 223–34.

Ellis, F.B., Mercer, E.R. & Milbourn, G.M. (1968) The contamination of grassland with radioactive strontium – II: Effect of lime and cultivation on the levels of strontium-90 in herbage. *Radiation Botany*, **8**, 269–84.

Ellis, F.B., Howells, H., Russell, R.S. & Templeton, W.L. (1960) Deposition of strontium-89 and strontium-90 on agricultural land and their entry into milk after the reactor accident at Windscale. UKAEA Report AHSB(RP)R2. HMSO, London.

English, R.E. (1979) Report of the technical assessment task force on chemistry. *The accident at Three Mile Island (Kemeny Report)*. Reports of the Technical Assessment Task Force. II, pp. 3–25. U.S. Govt. Printing Office, Washington D.C.

Farmer, F.R. & Beattie, J.R. (1976) Nuclear power reactors and the evaluation of population hazards. *Advances in Nuclear Science & Technology*, **9**, 1–72.

Forman, D., Cook-Mozaffari, P., Darby, S., Davey, G., Stratton, I., Doll, R. & Pike, M. (1987) Cancer near nuclear installations. *Nature*, **329**, 499–505.

Fredriksson, L., Garner, R.J. & Russell, R.S. (1966) Caesium-137. In: *Radioactivity and Human Diet*, ed. R.S. Russell. Pergamon, Oxford, pp. 319–53.

Fulker, M.J. (1987) Aspects of environmental monitoring by British Nuclear Fuels plc following the Chernobyl reactor accident. *Journal of Environmental Radioactivity*, **5**, 235–44.

Garner, R.J. (1960) An assessment of the quantities of fission products likely to be found in milk in the event of aerial contamination of agricultural land. *Nature*, **186**, 1063–4.

(1971) Transfer of radioactive materials from the terrestrial environment to animals and men. *C.R.C. Critical Reviews of Environmental Control*, **2**, 337–85.

Glasstone, S. & Dolan, P.J. (1977) *The Effect of Nuclear Weapons* (3rd edn). U.S. Dept. of Defence. N.T.I.S. Springfield, Va.

Graustein, W.C. & Turekian, K.K. (1986) ^{210}Pb and ^{137}Cs in air and soils measure the rate and vertical profile of aerosol scavenging. *Journal of Geophysical Research*, **91**, 14355–66.

Gudiksen, P.H., Harvey, T.F. & Lange, R. (1989) Chernobyl source term, atmospheric dispersion and dose estimation. *Health Physics*, **57**, 697–706.

Hardy, E., Krey, P., Klusek, C., Miller, K., Helfor, I., Sanderson, C. & Rivera, W. (1986) Observations and sampling by EML in Sweden, with preliminary gamma ray spectrometric data. In: *Report EML* 460 (ed. H.L. Volchok). U.S. Dept. of Energy, NTIS, Springfield, Va.

Hicks, H.G. (1982) Calculation of the concentration of any radionuclide deposited on the ground by offsite fallout from a nuclear detonation. *Health Physics*, **46**, 585–610.

Hoffman, F.O., Garton, C.T., Huckabee, J.W. & Lucas, D.M. (1982) Interception and retention of technetium by vegetation and soil. *Journal of Environmental Quality*, **11**, 134–40.

Hoffman, F.O. *et al.* (1989) Pasture grass interception and retention of ^{131}I, ^{7}Be and insoluble microspheres deposited in rain. Oak Ridge National Laboratory Report ORNL 6542, Springfield, Va. N.T.I.S.

Horan, J.R. & Gammil, W.P. (1963) The health physics aspects of the SL-1 accident. *Health Physics*, **9**, 177–86.

Hötzl, H., Rosner, G. & Winkler, R. (1989) Long term behaviour of Chernobyl fallout in air and precipitation. *Journal of Environmental Radioactivity*, **10**, 157–71.

Izrael, Y. (1989) Article in Pravda, summarised in 'Soviet data made public'. *Nature*, **338**, 367.

Jackson, D., Jones, S.R., Fulker, M.J. & Coverdale, N.G.M. (1987) Environmental monitoring in the vicinity of Sellafield following the deposition of radioactivity from the Chernobyl accident. *Journal of the Society for Radiological Protection*, **7**, 75–87.

Jaworowski, Z. & Kownacka, L. (1988) Tropospheric and stratospheric distributions of radioactive iodine and caesium after the Chernobyl accident. *Journal of Environmental Radioactivity*, **6**, 145–50.

Jost, D.T., Gäggeler, H.W., Baltensparger, V., Zinder, B. & Haller, P. (1986) Chernobyl fallout in size-fractionated aerosol. *Nature*, **324**, 22–3.

Kang, C. (1989) Measurements of whole body radiocaesium at the Harwell Laboratory, 1976–1988. *Health Physics*, **57**, 995–1001.

Krey, P.W., Klusek, C.S., Sanderson, C., Miller, K. & Helfer, I. (1986) Radiochemical characterisation of Chernobyl fallout in Europe. In: *Report EML 460* (ed. H.L. Volchok). U.S. Dept. of Energy. N.T.I.S. Springfield, Va. pp. 155–218.

Larsen, R.J., Sanderson, C.G. & Rivera, W. & Zamichieli, M. (1986) The characterisation of radionuclides in North American and Hawaiian surface air and deposition following the Chernobyl accident. In: *Report EML 640* (ed. H.L. Volchok) U.S. Dept. of Energy. N.T.I.S. Springfield, Va. pp. 1–104.

Linsley, G.S., Crick, M.J., Simmonds, J.R., & Haywood, S.M. (1986) Derived emergency reference levels. *National Radiological Protection Board Report DL10*. Chilton, Oxon.

Lockhart, L.,B., Patterson, R.L. & Saunders, A.W. (1965) Distribution of airborne activity with particle size. In: *Radioactive Fallout from Nuclear Weapon Tests*, ed. A.W. Klement Jr. CONF 765. N.T.I.S. Springfield, Va.

Loutit, J.R., Marley, W.G. & Russell, R.S. (1960) The nuclear reactor accident at Windscale, 1957: environmental aspects. In: *The Hazards to Man of Nuclear and Allied Radiations*. Cmnd 1235. HMSO, London.

Martin, C.J., Heaton, B. & Robb, J.D. (1988) Studies of ^{131}I, ^{137}Cs and ^{103}Ru in milk, meat and vegetables following the Chernobyl accident. *Journal of Environmental Radioactivity*, **6**, 247–59.

Medical Research Council (1959) Maximum permissible dietary contamination after the accidental release of radioactive material from a nuclear reactor. *British Medical Journal*, i, 967–9.

Medvedev, Z.A. (1976) Evidence on the Urals incident. *New Scientist*, **72**, 264.

(1979) *Nuclear Disaster in the Urals*. New York: Norton.

Megaw, W.J., Chadwick, R.C., Wells, A.C. & Bridges, J.E. (1961) The oxidation and release of iodine-131 from uranium slugs oxidising in air and carbon dioxide. *Reactor Science & Technology*, **15**, 176–84.

Middleton, L.J. (1959) Radioactive strontium and caesium in the edible parts of crop plants after foliar contamination. *International Journal of Radiation Botany*, **1**, 387–402.

Milbourn, G.M. & Taylor, R. (1965) The contamination of grassland with radioactive strontium – I, Initial retention and loss. *Radiation Botany*, **5**, 337–47.

Millard, G.C., Fraley, L. & Markham, O.D. (1983) Deposition and retention of ^{141}Ce and ^{134}Cs aerosols on cool desert vegetation. *Health Physics*, **44**, 349–57.

Miller, C.W. (1980) An analysis of measured values for the fraction of a radioactive aerosol intercepted by vegetation. *Health Physics*, **38**, 705–12.

Moorby, J. & Squire, H.M. (1963) The loss of radioactive isotopes from the leaves of plants in dry conditions. *Radiation Botany*, **3**, 163–7.

Mossop, I.A. (1960) Filtration of the gaseous effluent of an air-cooled reactor. *British Chemical Engineering*, **5**, 420–6.

Ng, Y.C. (1982) A review of transfer factors for assessing the dose from radionuclides in agricultural produce. *Nuclear Safety*, **23**, 57–71.

Nikipelov, B.V., Romanov, G.N., Buldakov, L.A., Babaev, N.S., Kholina, Y.B. & Mikerin, E.I. (1989) Accident in the Southern Urals on 29 September 1957. Transl. and published by International Atomic Energy Agency, Vienna. INFCIRC/368.

Osborne, M.F., Collins, T.L., Lorenz, R.A. & Strain, R.V. (1986) Fission product release and fuel behaviour in tests of LWR fuel under accident conditions. In: *Source Term Evaluation for Accident Conditions*, IAEA, Vienna, pp. 89–104.

Osburn, W.S. (1966) Ecological concentration of nuclear fallout in a Colorado mountain watershed. In: *Radiological Concentration Processes*, ed. B. Aberg & F.P. Hungate, Pergamon, Oxford, pp. 675–709.

Peirson, D.H. & Keane, J.R. (1962) Characteristics of early fallout from the Russian nuclear explosions of 1961. *Nature*, **196**, 801–7.

Peirson, D.H. & Cambray, R.S. (1965) Fission product fallout from the nuclear explosions of 1961 and 1962. *Nature*, **205**, 433–40.

Peirson, D.H., Cambray, R.S. & Spicer, G.S. (1966) Lead-210 and polonium-210 in the atmosphere. *Tellus*, **18**, 427–33.

Penney, W. (1957) *Accident at Windscale No. 1 Pile on 10th October 1957*. Cmnd 302, HMSO, London.

Persson, C., Rodhe, H. & de Geer, L.E. (1987) The Chernobyl accident – a meteorological analysis of how radionuclides reached and were deposited in Sweden. *Ambio*, **16**, 20–31.

Peters, L.N. & Witherspoon, J.P. (1972) Retention of 44-88μ simulated fallout particles by grasses. *Health Physics*, **22**, 261–6.

Pinder, J.E., Ciravolo, T.G. & Bowling, J.W. (1987) The interrelationships among plant biomass, plant surface area and the interception of particulate deposition by corn. *Health Physics*, **55**, 51–8.

Pinder, J.E., McLeod, K.W., Simmonds, J.R. & Linsley, G.S. (1985) Normalised specific activities for Pu deposition onto foliage. *Health Physics*, **49**, 1280–3.

Postma, A.K. & Pasedag, W.F. (1986) Overview of fission product release and the effectiveness of engineered safety features. In: *Source Term Evaluation for Accident Conditions*, IAEA, Vienna, pp. 621–32.

Roed, J. (1987) Dry deposition in rural and in urban areas in Denmark. *Radiation Protection Dosimetry*, **21**, 33–6.

Romney, E.M., Lindberg, R.G., Hawthorne, H.A., Bystrom, B.G. & Larson, K.H. (1963) Contamination of plant foliage with radioactive fallout. *Ecology*, **44**, 343–9.

Russell, R.S. (1965) Interception and retention of airborne material on plants. *Health Physics*, **11**, 1305–15.

(ed.) (1966) *Radioactivity and Human Diet*. Pergamon, Oxford.

Russell, R.S. & Possingham, J.V. (1961) Physical characteristics of fallout and its retention on herbage. *Progress in Nuclear Energy VI*, **3**, 2–26.

Sandalls, F.J., Eggleton, A.E.J. & Gaudern, S.L. (1989) Uptake of radiocaesium by upland vegetation in relation to soil type. Harwell Laboratory Oxfordshire Report AERE 13389.

Sansom, B.F. (1966) The metabolism of caesium-137 in dairy cows. *Journal of Agricultural Science*, **66**, 389–93.

Simmonds, J.R. & Linsley, G.S. (1982) Parameters for modelling the interception and retention of deposits from atmosphere by grain and leafy vegetables. *Health Physics*, **43**, 679–81.

Slinn, W.G.N. (1977) Some approximations for the wet and dry removal of particles and gases from the atmosphere. *Water, Air & Soil Pollution*, **7**, 513–43.

Smith, F.B. & Clark, M.J. (1986) Radionuclide deposition from the Chernobyl cloud. *Nature*, **322**, 690–1.

Squire, H.M. (1966) Long term studies of strontium-90 in soils and pastures. *Radiation Botany*, **6**, 49–67.

Stather, J.W., Wrixon, A.D. & Simmonds, J.R. (1984) The risks of leukaemia and other cancers in Seascale from radiation exposure. National Radiological Protection Board Report R-171, HMSO, London.

Stather, J.W., Dionian, J., Brown, J., Fell, T.P. & Muirhead, C.R. (1986) Addendum to Report R-171, HMSO, London.

Stewart, N.G. & Crooks, R.N. (1958) Long range travel of the radioactive cloud from the accident at Windscale. *Nature*, **182**, 627–30.

Stewart, N.G., Crooks, R.,N. & Fisher, E.M.R. (1961) Measurements of the radioactivity of the cloud from the accident at Windscale: data submitted to the I.G.Y. AERE Report M-857, Harwell, Oxon.

Trabalka, J.R., Egman, D. & Auerbach, S.I. (1980) Analysis of the 1957–1958 Soviet nuclear accident. *Science*, **209**, 345–53.

United Nations Scientific Committee on Effects of Atomic Radiations (1962, 1967). *Ionizing Radiation: Sources and Biological Effects*. New York, United Nations.

U.S. Nuclear Regulatory Commission (1981) Technical basis for estimating fission product behaviour during LWR accidents. Report NUREG-0772, Washington, D.C.

U.S.S.R. State Committee on the Utilization of Atomic Energy (1986) The accident at the Chernobyl nuclear power plant and its consequences. Safety Series No. 75-INSAG-I. IAEA, Vienna.

Van den Hoek, J., Kirchmann, R.J., Colard, J. & Sprietsma, J.E. (1969) Importance of some methods of pasture feeding, of pasture type and of seasonal factors on ^{85}Sr and ^{134}Cs transfer from grass to milk. *Health Physics*, **17**, 691–700.

Walton, G.N. (1961) Fission and fission products. In: *Atomic Energy Waste*, ed. E. Glueckauf. Butterworths, London, pp. 3–98.

Ward, G.M., Keszthelyi, B., Kangar, B., Kralovansky, V.P. & Johnson, J.E. (1989). Transfer of ^{137}Cs to milk and meat in Hungary from Chernobyl fallout. *Health Physics*, **57**, 587–92.

Webb, G.A.M., Simmonds, J.R. & Wilkins, B.T. (1986) Radiation levels in Eastern Europe. *Nature*, **321**, 821–2.

Winkelmann, I. *et al.* (1987) *Radioactivity Measurements in the Federal Republic of Germany after the Chernobyl Accident*. Institut für Strahlenhygiene des Bundesgesundheitamtes, Neuherberg, Munich.

Witherspoon, J.P. & Taylor, F.G. (1970) Interception and retention of a simulated fallout by agricultural plants. *Health Physics*, **19**, 493–9.

(1971) Retention of 1-44μ simulated fallout particles by soybean and sorghum plants. *Health Physics*, **21**, 673–7.

3

○ ○ ○ ○ ○ ○ ○ ○ ○ ○ ○ ○ ○ ○ ○ ○ ○ ○ ○

Radioiodine

3.1   Formation in fission and release to atmosphere

All the 24 isotopes of iodine with mass numbers 117 to 140 inclusive are formed in fission, either directly or by decay of a tellurium precursor. The fission chains including the four most important isotopes are shown in Fig. 3.1, with the percentage yields in slow neutron fission of ^{235}U, as given by Crouch (1977). Somewhat higher yields of the iodine isotopes are found in fission of ^{235}U, ^{238}U and ^{239}Pu by fast neutrons. Holland (1963) suggested that the fission yield of ^{131}I in nuclear weapons should be taken as 4%, but UNSCEAR (1982) give a yield of 2.9% for weapons, which is almost the same as the yield in slow fission of ^{235}U.

The activities of ^{131}I per MW thermal power in reactors, and per kT explosion in weapons, using the UNSCEAR yields, are shown in Table 2.2. Iodine is volatile and its precursor tellurium is relatively volatile, especially when in an oxidised state. In nuclear explosions, particles are formed as the fireball cools, and the more volatile fission products condense on them (Section 2.3). The isotopes of iodine, mass numbers 131 to 134, are formed by decay of tellurium isotopes which condense in this way, so iodine is probably initially particulate, but may become desorbed. During a period of weapons testing in 1961, more than 70% of ^{131}I in air at Harwell was particulate (Peirson & Keane, 1962). At Richland, Washington, Perkins (1963) found the gaseous fraction varying from 10 to 90%. Very little of the gaseous ^{131}I was in elemental or HI form.

Following near-ground explosions, iodine isotopes are concentrated in the smaller particles, but H-tests at high altitude produce very small particles, which equilibrate with other aerosols in the stratosphere, and there is little fractionation. After the Chinese test of 15.10.80, high

115

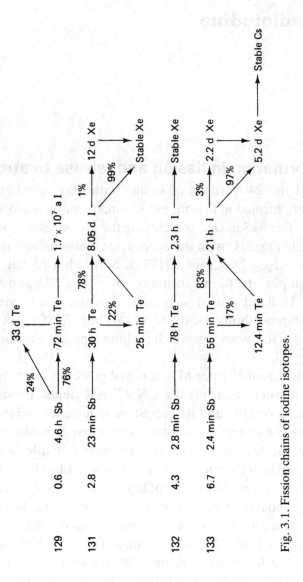

Fig. 3.1. Fission chains of iodine isotopes.

volume filters at Harwell showed a ratio of activities $^{131}I/^{140}Ba$ of 0.85 (corrected for decay to the date of the explosion) compared with the ratio 0.88 expected in fission of megaton weapons (Cambray, 1981).

The release of ^{131}I and other fission products in reactor accidents has been considered in the previous chapter. In the Windscale accident, the temperature in the fire zone reached an estimated 1300°C and 8 tonne of uranium metal melted. Over 25% of the ^{131}I in the melted fuel escaped to atmosphere. In the Chernobyl accident, the fuel was UO_2, the temperature exceeded 2000°C, and about 25% of the total reactor inventory of ^{131}I was released to atmosphere, as vapour or particulate aerosol. In the Three Mile Island accident, ^{131}I remained almost completely in the reactor coolant. The activities of ^{131}I released in reactor accidents, including that at Chernobyl, have totalled much less than the activities released from weapons tests (Table 2.3).

When reactor fuel is reprocessed, the fuel is dissolved in an oxidising environment, and iodine is liable to be released as vapour. Normally, several months elapse before fuel is reprocessed, and this minimises release of ^{131}I. Individual releases of about 4 TBq of ^{131}I occurred during the early operation of the Hanford reprocessing plant (Parker, 1956), and there was a release of 5.6 TBq from the Savannah River Plant in 1961, when fuel was reprocessed after a shorter than usual period of storage (Marter, 1963). These releases of ^{131}I, and those from the Sellafield reprocessing plant, are of only local significance compared with the releases in reactor accidents and weapon tests.

The activity of ^{129}I in reactors is very small compared with that of ^{131}I, amounting to about 3×10^4 Bq per megawatt-day of reactor operation (Soldat, 1976), but its half-life is very long (1.6×10^7 a) and its activity in the environment will increase if releases continue for many years. Since reprocessing began at the Karlsruhe plant in 1971, an estimated 7×10^9 Bq of ^{129}I have been released to the environment (Robens & Aumann, 1988). By comparing measurements of ^{129}I in soil near the Sellafield plant (Stewart & Wilkins, 1985) with similar measurements near Karlsruhe (Robens et al., 1989) it can be estimated that the cumulative release from Sellafield has been about 10-fold greater. These releases of ^{129}I are of no radiological consequence at present.

3.2   Characterisation of radioiodine

When experimental work with ^{131}I in the environment began at Harwell in 1949, it was thought appropriate to release the activity as elemental iodine, and to use NaOH bubblers, with added KI in solution, to sample the airborne activity (Chamberlain & Chadwick, 1953). When the Dido and Pluto research reactors were built at Harwell, they were provided with containment shells, and these were fitted with Na_2CO_3 scrubbers through which the air in the shell could be recirculated. It was soon found that penetrating forms of radioiodine were generated, which were not well absorbed in alkaline solution. To obtain better samples, F.G. May devised the May Pack (Megaw & May, 1962) in which air was drawn successively through a membrane filter to trap particles, two charcoal-loaded filter papers to trap elemental iodine vapour and a pack of granulated activated charcoal to trap organic iodine.

The May Pack has been used world-wide to characterise radioiodine. In some applications, glass fibre-filters have replaced membrane filters and copper or silver gauzes have been used to trap elemental iodine. The separation of iodine species in the May Pack is at best qualitative. Some inorganic iodine vapour may be adsorbed on the particulate filter. Conversely, iodine adsorbed on particles, and trapped on the particulate filter, may be desorbed during extended periods of sampling.

When, in 1960, it was decided to test the scrubbers fitted to the Dido and Pluto reactors, this was done by releasing ^{132}I into the containment shell during a reactor shut down. The ^{132}I was released as elemental vapour by heating crystalline iodine in a flow of nitrogen. Varying amounts of stable iodine carrier were used, giving initial air concentrations ranging from 0.013 to 0.14 $\mu g\,m^{-3}$ in different tests. May Packs were used to sample the air periodically up to 6 h from the release. In some experiments, additional stable I_2 carrier (100 mg) was released some time after the start of the experiment (Megaw & May, 1962).

Figure 3.2 shows results of the experiment in which the initial concentration of stable I_2 was 0.013 $\mu g\,m^{-3}$. The scrubber was not in operation. All results are corrected for radioactive decay of ^{132}I. The airborne concentration rose initially as the source was mixed in the air within the containment shell. The activity on the charcoal-loaded papers was due mainly to inorganic iodine, and this declined with a half-life of about 30 min, as ^{132}I was adsorbed on surfaces. Plaques of various materials were exposed periodically to monitor the deposition. In other experiments with more iodine carrier, loss by deposition was more rapid.

The ratio of activities on the first to the second charcoal paper decreased with time from about 250 at 20 min to 25 at 4 h after the release, showing an increasing proportion of penetrating species. The activity on the membrane (particulate) filter peaked at 80 min and then declined slowly. The activity on the charcoal pack (organic iodine) increased throughout. When 100 mg of I_2 carrier were released at 4 h, about 50% of the previously deposited ^{132}I came back into the air, showing that the adsorption on surfaces was partly reversible.

The particulate activity trapped on the membrane filters in the sampling packs increased during the first hour (Fig. 3.2). The concentration of condensation nuclei in air in the reactor shell was 1.2×10^{10} m^{-3}. Megaw & May showed that an accommodation coefficient of 5×10^{-3} (compare Section 1.12) would explain the observed rate of increase in particulate iodine due to adsorption on the nuclei. The subsequent decline in particulate activity was due to deposition of nuclei on surfaces. Surprisingly, in this and other experiments, release of stable iodine vapour into the containment shell 4 h after the start of the experiment made little difference to the concentration of particulate ^{132}I. Subsequently, Clough *et al.* (1965) showed that the amount of

Fig. 3.2. Activity of radioiodine in air of reactor shell. ○, First charcoal paper; △, second charcoal paper; ×, membrane filter; □, charcoal pack.

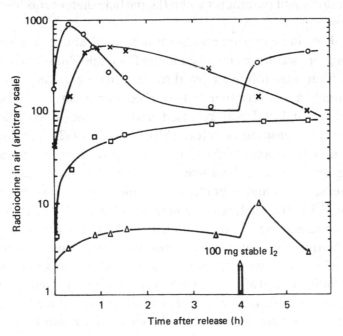

adsorption depends strongly on the nature of the nuclei. Oil smoke from vacuum pumps, which were used in the Dido shell, produce nuclei which are effective absorbents.

Further studies on the adsorption of radioiodine on particles were carried out in the laboratory by Garland (1967). Elemental iodine, labelled with ^{132}I, was mixed with room air in a 20-1 vessel to give vapour concentrations varying from 0.6 to 2400 μg m^{-3}. After a stay time of about 10 min, the vapour was drawn down a tube coated internally with activated charcoal, which removed both inorganic and organic iodine vapour. Particles passed down the tube and were trapped on membrane filters, which were found to hold 0.1 to 0.5% of the ^{132}I entering the tube, depending on the number and size of the condensation nuclei in the air and on the amount of stable iodine carrier.

From the results, Garland deduced that the accommodation coefficient for adsorption of ^{132}I on particles was in the range 2×10^{-3} to 10^{-2}, in agreement with the results of Megaw & May. There was some reduction of accommodation coefficient when the stable iodine concentration was increased. Most of the adsorption was on submicrometre particles, but experiments in which ^{132}I and ^{212}Pb were both adsorbed from the vapour phase showed a smaller proportion of ^{132}I than ^{212}Pb on the smallest particles. This is to be expected from the equation (1.18) governing adsorption on particles, in which the accommodation coefficient is the dominant parameter when the particle diameter is less than about 0.1 μm.

As a result of the experimental work described above, a modified decontamination circuit was installed in the Dido and Pluto reactors. A particulate filter was followed by three alternative iodine removal systems, namely a wet scrubber, a bed of copper mesh, and a bed of charcoal. Elemental ^{132}I was released and airborne activity in the containment was measured as before (Morris *et al.*, 1962). In addition, the decontamination factor (*DF*) of the clean-up systems was measured by sampling upstream and downstream of them. The *DF* values of the filter plus wet scrubber and filter plus copper mesh systems were similar. Initially the ^{132}I in the shell was elemental and the *DF* was about 20. After 3 h, the remaining ^{132}I was nonelemental and the *DF* only about 1.5, since neither the wet scrubber nor the copper mesh removed nonelemental iodine. The filter plus charcoal system showed an initial *DF* of 2000 falling to 100 after 3 h. Clearly, a recirculating system which removed only inorganic iodine was of little advantage, since I$_2$ was lost by adsorption on surfaces faster than the clean-up system treated the air. Decontamination systems including particulate filters with charcoal beds were installed in the AGR reactors.

3.3   Identification of penetrating forms of radioiodine

Only limited progress has been made in identifying the mechanisms by which gaseous nonelemental iodine is released from irradiated fuel or formed in containments or in the atmosphere. Eggleton & Atkins (1964) generated radioiodine by oxidation of $Na^{131}I$ or irradiated uranium or by heating irradiated UO_2 in CO_2. Elemental and particulate iodine were removed, and the remaining ^{131}I was passed either into an H_2SO_4 bubbler or into a liquid oxygen cold trap. The activity was found to comprise two fractions, A and B. Fraction B partitioned equally between benzene and M/50 sulphuric acid but was not identified. Fraction A was readily soluble in benzene and was identified as CH_3I by a determination of molecular weight and by gas chromatography.

In further experiments (Collins & Eggleton, 1965), miniature irradiated UO_2 fuel elements were heated to 1500°C in either CO_2/CO or CO_2/air mixtures. Between 30 and 50% of I, Te and Cs fission products were released, but less than 0.2% of Sr. In CO_2/air the released iodine was mainly elemental (as determined by adsorption on copper gauzes in May Packs). In CO_2/CO, iodine was released as a mixture of alkyl iodides, with methyl iodide dominant. The mode of formation of alkyl iodides in these experiments remains uncertain.

In normal operation of AGR reactors, radioiodine can be detected in the circulating coolant. When samples of coolant were withdrawn down the long narrow steel pipes of the burst-can-detection (BCD) system, most of the iodine was found to be in a non-reactive form, probably methyl iodide. More recently, samples taken through a wide-bore access system at the Hinkley Point reactor have shown the reactive component exceeding the CH_3I component by a factor of 10 (Hood & Clough, 1984). When experimental injections of ^{131}I as CH_3I were made into the coolant of AGR reactors, the activity disappeared very rapidly by plate-out in the circuit, with an initial half-life as short as 12 s (Garland *et al.*, 1984). Hood & Clough (1984) explained this behaviour by postulating that CH_3I is decomposed by reaction with radicals derived by radiolysis of CO_2. These reactions lead to atomic iodine which then plates out.

A case can be made out on thermodynamic grounds that fission product iodine should react with fission product caesium in oxide fuel, and iodine should be released as CsI. The releases of ^{131}I and ^{137}Cs in reactor accidents, as fractions of the inventory, have been similar (Tables 2.4–2.6), but differences have been observed in their subsequent behaviour. It is possible that CsI has been deposited in the high-temperature region of experiment rigs, and has subsequently undergone decomposition.

A volatile iodine species, neither elemental nor organic, which has been found in steam/air atmospheres, has been identified as hypoiodous acid (HOI) (Cartan *et al.*, 1968). In water-cooled power reactors, any fission products released from fuel will pass into hot alkaline water and thence to a steam–air mixture. These conditions are thought to favour the formation of HOI (Keller *et al.*, 1970), but the evidence is indirect. For example, tests for elemental iodine or iodine with an oxidation state higher than that of HOI gave negative results.

To monitor radioiodine species in gaseous effluent from water-cooled reactors, a four-component iodine sampler was used (Voil-lequé, 1979), designed to separate iodine species as follows:

High efficiency filter: Particulate
Cadmium iodide on Chromosorb B: Elemental
4-Iodophenol on alumina: Hypoiodous acid
Charcoal or silver zeolite: Organic.

Hypoiodous acid is reported to be the sole iodine species capable of the iodination of phenols (Cofman, 1919). The iodophenol bed was over 90% efficient for trapping HOI (prepared by sparging from a dilute alkaline iodine solution) and less than 1% efficient for CH_3I. Voillequé (1979) used this sampler to analyse effluents from Boiling Water Reactors and found ^{131}I to be 11% particulate, 38% I_2, 29% HOI and 22% organic.

The chemical form of ^{131}I released from the Hanford reprocessing plant was studied by Haller & Perkins (1967). Using an NaOH scrubber followed by a charcoal trap to sample the effluent, they found that 40% was organic and 60% inorganic. For further investigation, samples were collected in liquid nitrogen. After removal of inorganic iodides, the organic iodine compounds were identified by gas chromatography. Methyl iodide was most abundant, but a wide spectrum of organic iodides, with boiling points ranging up to that of n-octyliodide (255°C) was found. Some of the compounds probably derived from iodination of the organic solvents used in the Hanford process.

Wershofen & Aumann (1989) used a filter system to sample [129]I and also stable [127]I near the Karlsruhe Reprocessing Plant. Polycarbonate filters, cellulose filters impregnated with tetrabutylammonium hydroxide and activated charcoal were used in series to estimate particulate inorganic gaseous and organic gaseous iodine respectively. About 60% of both [129]I and [127]I in air were found to be organic, with the remainder divided equally between the particulate and inorganic gaseous phases.

3.4   Environmental sampling of radioiodine

May Packs were not available to sample the ^{131}I emitted in the Windscale accident of October 1957. Use was made of filters operated by Local Authorities in most large towns to measure smoke. These were mostly paper filters operated at low flows of about 30 l min^{-1}. They were collected after the accident from several European countries as well as the UK, and analysed at Harwell for ^{131}I, ^{137}Cs and [210]Po (Stewart & Crooks, 1958; Stewart *et al.*, 1961). The physicochemical form of the ^{131}I, and the sampling efficiency of the filters are not known. By reference to measurements of ^{133}Xe in air (Section 2.6), Chamberlain (1981) deduced that the dosage of ^{131}I in the London area, as measured by the filters, was about 20% of what would have been found if all the ^{131}I in the melted fuel had escaped from the reactor and none had been lost by deposition. The sampling filters must therefore have been reasonably efficient for inorganic iodine.

When the Chernobyl accident occurred, better sampling systems were available. Winkelmann *et al.* (1987), at Neuherberg, Munich, used a sampling unit comprising aerosol filter, molecular sieve and charcoal pack. The radioiodine trapped in the three components was assumed to be particulate (I_p), inorganic (I_i) and organic (I_o). Any HOI would be included in the organic fraction. Table 3.1 shows the percentage distribution of the dosages found. The percentage of I_o increased from 43% in the first 3 d of the sampling period to 59% in the last 7 d. The distribution of [133]I in the earlier period was similar to that of [131]I. During the later period, the activity of [133]I was below the limit of detection.

The distribution of [132]I between the sampling components was different. Its radioactive half-life is short, and the activity on the sampler derived from its parent isotope [132]Te (Fig. 3.1). The distribution of [132]Te (Table 3.1) shows that it was almost entirely particulate. Hence the 30% of [132]I found on the molecular sieve, and 5% on the charcoal pack must have derived from [132]I which desorbed from the particles, either in the atmosphere or after capture on the filter.

Elsewhere, two-component samplers were operated to separate the particulate and gaseous (I_g) fractions of iodine from Chernobyl, without attempting to distinguish inorganic and organic gaseous forms. The I_p/I_g ratio was about 0.25 in Scandinavia (Devell *et al.*, 1986), 0.2 at Harwell (Cambray *et al.*, 1987), 0.5 in Japan (Aoyama *et al.*, 1986) and 0.5 in New York (Leifer *et al.*, 1986). The particulate fraction may have been derived partly from discrete particles of fuel disseminated in the accident, but it is noteworthy that the distribution between I_p, I_i and I_o observed at Munich is not dissimilar to the distribution of stable iodine as reviewed in the next section.

If during the operation of a May Pack sampler, iodine is desorbed from particles on the membrane filter, as demonstrated in laboratory experiments by Garland (1967), this will have the effect of reducing the apparent I_p/I_g ratio. In Winkelmann *et al.*'s (1987) series of Munich samples, covering the period from 3 to 13 d from the start of the emission from Chernobyl, there was an apparent decline in I_p/I_g with time. This may have been a genuine effect, or may have been because the later sampling periods were of longer duration. In Japan, Aoyama *et al.* (1986), sampling Chernobyl iodine daily, found I_p/I_g about 0.5, whereas Noguchi & Murata (1988), with samplers run for 3 to 4 d, found I_p/I_g averaging 0.2.

Noguchi & Murata modified the May Pack to include silver wire meshes to catch inorganic iodine downstream as well as upstream of the particulate filter. Activity on the second meshes was ascribed to desorption from particles. However, it is not certain that iodine desorbed from particles, or other surfaces, is necessarily inorganic. In the experiments

Table 3.1. *Distribution of radioiodine from Chernobyl in Munich Sampler*

Isotope	Half-life	Per cent		
		I_p	I_i	I_o
^{131}I (*a*)	8.0 d	28	29	43
^{131}I (*b*)	8.0 d	17	24	59
^{132}I (*a*)	2.3 h	64	30	5
^{133}I (*a*)	21 h	20	32	48
^{132}Te (*a*)	3.2 d	98	1	1

(*a*) Samples from 29/4 to 1/5/86.
(*b*) Samples from 2/5 to 8/5/86.

in the Dido shell (Section 3.2) inorganic iodine was released. Organic iodine was generated, either in the gas phase or on surfaces, and the latter appears more probable.

3.5   Particulate and gaseous stable iodine

Table 3.2 shows measurements of particulate and gaseous stable iodine in the atmosphere. Moyers *et al.* (1971) used membrane filters and activated charcoal to collect the particulate and gaseous fractions in air at Boston, and found the ratio I_p/I_g to be correlated with the concentration of particulate lead in the air. It was not inferred that I_p was combined with particles of lead, but rather that the concentrations of lead served as an index of the total airborne particulate. Moyers *et al.* expressed their results as

$$I_p/I_g = (1800 \pm 600)M + 0.16 \tag{3.1}$$

where M (g m^{-3}) is the total particulate loading. The slope of this regression is similar to a relation derived by Clough *et al.* (1965) from laboratory experiments on the adsorption of ^{131}I on particles.

Rahn *et al.* (1976) interposed a cellulose filter impregnated with LiOH between membrane filter and charcoal pack to separate the inorganic and organic fractions of gaseous iodine, with results shown in the last four rows of Table 3.2. The average distribution of stable iodine in Rahn *et al.*'s results was I_p 8%, I_i 34%, I_o 58%. The ratio I_i/I_o was similar to the corresponding ratios for ^{131}I and ^{133}I in Table 3.1, but a smaller proportion of the stable iodine was particulate. Rahn *et al.* ran their samplers for 2 to 4 d, whereas Winkelmann *et al.* changed them after 4 to 18 h.

It is believed that CH_3I is produced by marine algae and released from seawater, and that this constitutes the main source of stable iodine in the atmosphere (Lovelock *et al.*, 1973). Elemental iodine may be liberated from the sea surface by ultraviolet light (Miyake & Tsunogai, 1963), or by the action of ozone (Garland & Curtis, 1981), but I_2 is dissociated very rapidly by photochemical action, and its mean residence time in daylight air is less than a minute.

CH_3I is also dissociated, but with a residence time of about a week (Chameides & Davis, 1980). The I atoms formed by photolysis may form compounds such as IO, HOI or $IONO_2$ or recombine as I_2. Jenkin *et al.* (1985) developed a numerical model of the behaviour of iodine, and concluded that in daylight and in the presence of a typical concentration of NO_2 (2 ppb), 90% of iodine would be present as $IONO_2$.

Table 3.2. *Stable iodine in the atmosphere*

| | Concentration (ng m^{-3}) | | | | Ratio | |
Location	Particulate I_p	Inorganic I_i	Organic I_o	Total gaseous I_g	I_p/I_g	Reference
Boston	6.8	n.a.	n.a.	14	0.49	Moyers et al. (1971)
Hawaii	2.4	n.a.	n.a.	8.3	0.29	Moyers & Duce (1972)
Antarctica	0.7	n.a.	n.a.	2.4	0.29	Duce et al. (1973)
N.W. Territories	0.2	0.4	3.0	3.4	0.06	Rahn et al. (1976)
Bermuda	3.8	17	28	45	0.08	Rahn et al. (1976)
Arizona	1.3	11	5	16	0.08	Rahn et al. (1976)
Kansas	2.4	5	12	17	0.14	Rahn et al. (1976)

n.a. Not applicable.

It is difficult to summarise effectively the information on the chemical state of radioactive or stable iodine in the atmosphere. Most samples are found to contain particulate, gaseous inorganic and gaseous organic forms. The processes of adsorption of elemental iodine on particles are reasonably well understood, but no good experimental data are available to show whether the organic fraction of fission product iodine is formed at source, by gas-phase reactions, in air, or by reactions at surfaces. Surfaces where reactions might occur include the walls of containment vessels, the surfaces of particles and possibly the filters used to catch particulate matter in sampling trains.

3.6   Deposition of radioiodine to surfaces

For hazard evaluation, the important properties of radioiodine are its reaction rates at surfaces, particularly the uptake in the lung and the deposition to grass. The uptake in the lung may not depend critically on the physicochemical state. Morgan *et al.* (1966) found that methyl iodide, labelled with ^{131}I, was readily absorbed in the lungs of volunteer subjects. Inorganic iodine, whether gaseous or adsorbed on particles of submicrometric size, can also be expected to be absorbed in the lung. The physicochemical form of iodine is critical to the rate of deposition on grass and entry thence into milk. These aspects are considered in the next section.

Elemental iodine is a reactive gas, and the rate of uptake on certain surfaces is controlled by the rate of diffusion through the boundary layer over the surface. At some surfaces, tracer quantities of iodine are adsorbed irreversibly, at others reversibly. In most applications the amount of iodine on the surface is much less than a monolayer, and the equilibration between the adsorbed and airborne iodine cannot be considered in terms of vapour pressure. In 1949, experiments were started at Harwell, both in the wind tunnel and in the field, to study the deposition of elemental ^{131}I vapour to surfaces.

The velocity of deposition of radioiodine, defined as in equations (2.3) and (2.4), is controlled by the rate of adsorption at the surface. When the adsorption or chemisorption at the surface is strong, the rate of deposition of ^{131}I, when expressed in terms of suitable non-dimensional parameters, is similar to the rate of heat transfer to or from the surface (Chamberlain, 1953). In field experiments, artificial leaves of copper or silver foil were attached to vegetation, and it was found that the uptake of ^{131}I to real leaves was typically 0.4 to 0.7 times the uptake to artificial leaves. Since a thin film of paraffin was found to inhibit sorption of ^{131}I on copper, it was thought likely that the waxy cuticle on leaves acted similarly, and that uptake of ^{131}I was through the pores (stomata) which control transport of water vapour from, and transport of CO_2 to, the leaf.

Laboratory experiments (Barry & Chamberlain, 1963; Adams & Voillequé, 1971; Guenot *et al.*, 1982; Garland & Cox, 1984) have shown that stomatal aperture does control the deposition of radioiodine when the relative humidity is low, but not when it is high. Guenot *et al.* (1982), and Garland & Cox (1984) measured the deposition of elemental iodine vapour to individual leaves of bean plants. The rate of transpiration of water from the leaves was also measured. Figure 3.3 shows the velocity of deposition to the bean leaves in relation to the relative humidity. In calculating v_g, the area of the leaf was the area of one side only. The variation in v_g to individual leaves has been used to calculate the one standard error bars in Fig. 3.3. The outwards flux of water vapour was also expressed as a velocity or conductance, by relating it to the difference in vapour pressure between the mesophyll of the leaf and ambient air. The ratios of conductances for I_2 and H_2O, from both sets

Fig. 3.3. Velocity of deposition of radioiodine vapour to bean leaves. In light: ○, Guenot *et al.*, 1982; △, Garland *et al.*, 1984. In dark: ▲, Garland *et al.*, 1984.

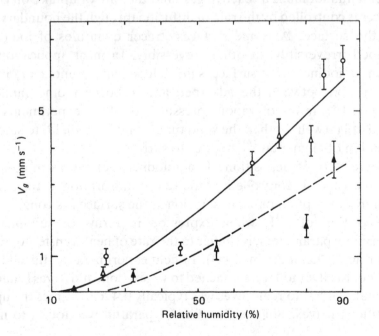

of experiments, are shown in Table 3.3. The molecular diffusivities of I_2 vapour and water vapour are 8×10^{-6} m^2 s^{-1} and 2.5×10^{-5} m^2 s^{-1} respectively, so a ratio of 0.32 would be expected if diffusion through the stomata is the controlling factor. At $RH < 50\%$, the ratio of conductances averaged 0.52 but at $RH > 50\%$ it was 2.3. Evidently, iodine was adsorbed on the cuticle of the leaf when the RH was high.

3.7   Deposition of iodine in the field

Table 3.4 summarises results of field experiments on the deposition of iodine vapour. The results are subject to variations in the method of sampling the grass. In the Harwell experiments (Chamberlain, 1960) the grass was cut as close to the ground as possible, and in the later experiments (Chamberlain & Chadwick, 1966) the underlying matt was included. In the Idaho experiments (Bunch, 1968), grass was grown in irrigated plots and was cut to 50 mm above ground. In the Jülich experiments (Heinemann & Vogt, 1980) herbage was cut at a height considered typical of grazing cattle.

The mean v_g as reported in all experiments with elemental iodine in Table 3.4 is 18 mm s^{-1}. If the leaf area index is 5, counting the area of one side of each leaf, this is consistent with a v_g to individual leaves of 3.6 mm s^{-1} (compare Fig. 3.3).

Elemental iodine, and probably also HI vapour, are strongly adsorbed by leaf surfaces, but this is not true of organic iodine. In a single experiment with radioiodine as CH_3I (Atkins et al., 1967), v_g was only 0.01 mm s^{-1}. Because uptake by vegetation is so restricted, the release of radioiodine as CH_3I does not cause significant contamination of herbage, but the activity is readily absorbed from the lung after inhalation by humans (Morgan et al., 1966), and presumably also by cattle, so ^{131}I as CH_3I could enter milk by that route.

Table 3.3. *Ratio of conductances of radioiodine and water vapour to or from bean leaves*

Relative humidity	No. of exps.	$v(I_2)/v(H_2O)$ Mean	S.E.
<50%	5	0.52	0.08
>50%	8	2.3	0.6

Table 3.4. *Deposition of radioiodine vapour in the field*

Source	Method of release	No. of exps.	Air samplers	Distance Source-samplers (m)	Surface	v_g (mm s^{-1}) (± s.e.)	v_D (m^3 kg^{-1} s^{-1}) (± s.e.)	Reference
$^{131}I_2$	Atomisation of soln. in CCl$_4$	7	NaOH bubblers	15 to 100	Grass	21 ± 3	0.18 ± 0.04	Chamberlain (1960)
$^{131}I_2$	Oxidation of iodide	8	Charcoal filter paper	50 to 100	Grass	17 ± 3	0.12 ± 0.02	Chamberlain & Chadwick (1966)
$^{131}I_2$	Oxidation of iodide, or volatilisation of soln. in CCl$_4$	11	Carbon trap	50 to 300	Grass & matt	10 ± 3	0.12 ± 0.04	Bunch (1968)
Stable I$_2$	Heating I$_2$ crystals	18	Charcoal filter paper	20	Grass	10 ± 2	0.08 ± 0.01	Heinemann & Vogt (1980)
Stable I$_2$	Heating I$_2$ crystals	5	Charcoal filter paper	20	Wet grass	27 ± 10	0.14 ± 0.03	Heinemann & Vogt (1980)
Stable I$_2$	Heating I$_2$ crystals	5	Charcoal filter paper	20	Clover	20 ± 6	0.16 ± 0.02	Heinemann & Vogt (1980)
CH$_3$I	As vapour from vacuum apparatus	1	Carbon pack	25	Grass	0.01		Atkins et al. (1967)

Table 3.4 also shows the velocity of deposition of iodine vapour normalised in respect of herbage density as:

$$v_D = \frac{\text{amount deposited per kg dry weight of herbage}}{\text{time integral of amount per m}^3} \qquad (3.2)$$

The units of v_D are $m^3 \, kg^{-1} \, s^{-1}$. It is less liable to sampling variation than v_g. The results of Table 3.4 give a mean v_D of $0.13 \, m^3 \, kg^{-1} \, s^{-1}$.

In the Idaho experiments, 5 mg of stable iodine carrier were added to the ^{131}I in the preparation of the source. There were no significant differences in v_D with source/sample distances up to 300 m. In the Jülich experiments, sources of 25 g I_2 were released, and the deposition was assessed from the increase in stable iodine on the herbage, at ranges of about 20 m, giving values of v_D not very different from those obtained in Idaho. This suggests that the predicted rapid photolysis of I_2 vapour does not affect its deposition greatly at source/sample distances of a few hundred metres.

Table 3.5 summarises data on the deposition of ^{131}I to herbage in operational and accidental conditions. Parker (1956) monitored ^{131}I in air and herbage near the Hanford Separation Plant and deduced a value 28 mm s^{-1} for v_g. This may have been relative to the inorganic fraction only of the airborne ^{131}I. Measurements of ^{131}I from the 1957 Windscale accident, the 1961 Idaho (SLI) accident, and the 1986 Chernobyl accident give values of v_g in the range 1–10 mm s^{-1}, calculated relative to the total activity of ^{131}I in air, not all of which was inorganic vapour.

The dry deposition of particulate radioiodine depends on the particle size. Peirson & Keane (1962) measured ^{131}I in air, rain and grass during a series of Russian nuclear tests in 1961, and deduced a dry deposition velocity of 2.6 mm s^{-1}. The ^{131}I was mainly particulate. If the herbage density was 0.2 kg m^{-2} dry weight (typical of the not very productive Harwell site), this gives $v_D = 0.013 \, m^3 \, kg^{-1} \, s^{-1}$. Voillequé & Keller (1981) derived a value 0.01 $m^3 \, kg^{-1} \, s^{-1}$ from measurements of long-range fallout on a Michigan farm in 1964/5.

Only limited information is available on the washout of radioiodine vapour by rain. Measurements of stable iodine in rain, compared with iodine vapour in air (Whitehead, 1984), give a value about 30 for the washout ratio W, defined as the ratio (I per kg rain)/(I per kg air). For a moderately heavy rainfall of 1 mm h^{-1}, this would imply a velocity of deposition in rain of 7 mm s^{-1}, which is the same order of magnitude as the velocity of dry deposition. The washout ratio of particulate fission products is typically about 500 (Table 2.12), so radioiodine deposited in

Table 3.5. *Deposition of iodine vapour in operational and accident conditions*

Source	Distance (km)	Surface	v_g (mm s^{-1})	v_D (m^3 kg^{-1} s^{-1})	Reference
Hanford Separation Plant	10–50	Grass	28	—	Parker (1956)
Savannah River Plant	1	Grass	—	0.06	Marter (1963)
SL1 accident	8.5	Sagebrush	2.5 ⎫	0.04	Islitzer (1962)
	67	Sagebrush	2.0 ⎬		
		Sagebrush	2.3 ⎭		
Windscale accident	160	Grass	3		Stewart & Crooks (1958)
	300	Grass	1		
Chernobyl accident	*ca.* 1000	Grass & soil	3–10		Devell *et al.* (1986)
	ca. 2000	Grass & soil	3		Cambray *et al.* (1987)

rain will derive mainly from the particulate fraction if this amounts to more than about 10% of the total airborne activity, as was the case after the Chernobyl accident.

3.8 Deposition near a reprocessing plant

Washout by rain becomes fully effective only when the cloud of activity extends up to the height of the rain-forming clouds. Within less than 10 km from a near-ground source, dry deposition is usually more effective than washout.

The cumulative fallout of ^{129}I in soil near the Karlsruhe reprocessing plant (Robens *et al.*, 1989) can be related to the ^{129}I activity released from the 60-m tall stack of the plant. If A_I is the deposit per m^2 of ground, and Q the total emission, the fractional deposition A_I/Q is equal to $v_g\chi$, where χ is the time-integrated concentration, or dosage, per unit emission. Bryant (1964) used meteorological parameters weighted according to the prevalence of different wind speeds and atmospheric stabilities to calculate χ as a function of distance and height of stack. Robens *et al.* took soil samples to the NE and ENE of the Karlsruhe plant, in the direction of the prevailing wind. In Fig. 3.4, Robens *et al.*'s results are compared with prediction, with v_g taken as 3 mm s^{-1}. A factor 2 has been applied, as suggested by Bryant, to allow

Fig. 3.4. Deposit of ^{129}I in soil near Karlsruhe Reprocessing Plant as fraction of total emission. Curve is theoretical with $v_g = 3$ mm s^{-1}.

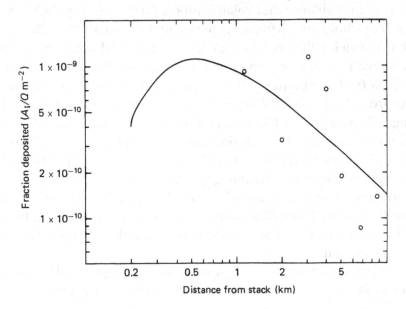

for the frequency of wind in the sector sampled compared with uniform swing of wind. The measured activities of ^{129}I were of the order expected from theory.

3.9   Retention of ^{131}I on herbage

Iodine is lost from herbage by the same processes which cause field loss of ^{90}Sr, ^{137}Cs and other nuclides (Section 2.13). There is also the possibility of revolatilisation of iodine. If λ_G is the rate constant of field loss (fraction of iodine per unit area of ground lost from vegetation per second) and λ_1 the rate constant of radioactive decay, the combined apparent or effective loss rate is $\lambda_E = \lambda_G + \lambda_1$. The effective half-life is $T_E = 0.693/\lambda_E$. The use of the term half-life implies that field loss is exponential and T_E invariant with time, which is not always true.

T_E has been measured by sequential sampling of vegetation after deposition of ^{131}I in field experiments, reactor accidents, emissions from separation plants and weapon tests. Chamberlain & Chadwick (1966) listed eight measurements in UK and USA giving T_E between 3.5 and 6 d. Figure 3.5 shows the activity of ^{131}I and ^{137}Cs on grass at Harwell after the Chernobyl accident. The field loss half-life of ^{137}Cs was 14 d ($\lambda_G = 0.05$ d^{-1}, compare Table 2.16). For ^{131}I, T_E was 4.7 d, giving $\lambda_E = 0.147$ d^{-1} and $\lambda_G = 0.147 - 0.087 = 0.06$ d^{-1}. Thus ^{131}I was lost from the grass slightly faster than ^{137}Cs. However, after the Windscale accident, it was the other way round, ^{137}Cs being lost slightly faster than ^{131}I (Booker, 1958).

There is evidence that volatilisation contributes only slightly to field loss of radioiodine. After an experimental release of ^{131}I, Chamberlain & Chadwick (1966) covered an area of grassland with a plastic cover, from within which air was drawn through an iodine trap. In 14 d, only 0.16% of the deposited ^{131}I was volatilised and recovered in the trap. Amiro & Johnston (1989) grew bean plants in nutrient solution containing ^{125}I. Activity was taken up into the leaves, and the rate of volatilisation was measured by enclosing them in cuvettes. The flux density, per unit plan area of leaf, was about 2×10^{-9} Bq m^{-2} s^{-1} per Bq kg^{-1} wet weight in the leaf. Assuming a wet weight of 0.25 kg m^{-2}, this corresponds to a loss rate of 8×10^{-9} s^{-1}, or 7×10^{-4} d^{-1}, which is small compared with observed rates of field loss. Both Chamberlain & Chadwick and Amiro & Johnston noted that the volatilised iodine was not elemental.

Volatilisation may be more significant when appreciable quantities of stable iodine carrier are present. Heinemann & Vogt (1980) measured iodine in grass on their experimental plots after each of their releases of the stable element. Exponential loss was found with $\lambda_G = 0.092 \pm$ (s.e.) 0.005 d^{-1}. If applied to ^{131}I, this gives $\lambda_E = 0.18$ d^{-1}, $T_E = 3.9$ d. In the sequel, $T_E = 5$ d, $\lambda_E = 0.14$ d^{-1} will be assumed for ^{131}I.

If a constant airborne concentration of ^{131}I is maintained until the concentration reaches equilibrium, the air/grass transfer factor defined:

$$\text{Transfer factor} = \frac{^{131}\text{I per kg dry wt. of herbage}}{^{131}\text{I per m}^3 \text{ of air}}$$

$$= v_D/\lambda_E \qquad\qquad (3.3)$$

Fig. 3.5. ^{137}Cs and ^{131}I on grass at Harwell as a function of time after deposition from Chernobyl (Cambray *et al.*, 1987).

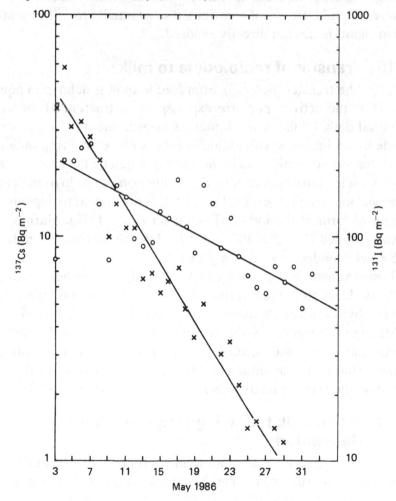

If $v_D = 0.1$ m^3 kg^{-1} s^{-1} and $\lambda_E = 0.14$ d$^{-1} = 1.6 \times 10^{-6}$ s^{-1}, then the transfer factor air/grass is 6×10^4 m^3 kg^{-1}. Somewhat lower values, in the range 2 to 4×10^4 m^3 kg^{-1} have been obtained by comparing measurements of ^{131}I in air and herbage during periods when activity has been emitted from chemical separation plants (Marter, 1963; Eisenbud & Wrenn, 1963).

Because its radioactive life is short, direct deposition on foliage is the only important route of entry of ^{131}I into plants, but ^{129}I is also taken up from the cumulative fallout in soil. Soldat (1976) calculated that the two routes would contribute equally to concentrations of ^{129}I in grass if fallout continued for 60 a. Robens *et al.* (1989) measured ^{129}I in soils and plants near the Karlsruhe reprocessing plant, which began operation in 1971. The results can be interpreted as showing that most of the ^{129}I in foliage directly exposed to fallout derived from direct deposition, whereas uptake from soil contributed substantially to ^{129}I in grain and other plant tissues not directly exposed.

3.10   Transfer of radioiodine to milk

The transfer factor F_m from feed to milk is defined in equation (2.11) as the activity per litre expressed as a fraction of the activity ingested daily by the cow. Numerous measurements of F_m have been made by dosing cows with iodine isotopes or by comparing radioactive or stable iodine in milk and in the herbage eaten by the cows. Table 3.6 shows a selection of the results. F_m depends on the yield of milk and on seasonal factors (Garner, 1971), but it does not appear to depend on the chemical form of the iodine (Bretthauer *et al.*, 1972). Garner (1971) recommended $F_m = 5 \times 10^{-3}$ d l^{-1} for UK conditions, but noted that US results tended towards a higher value.

Lower values, in the range 1 to 3×10^{-3} d l^{-1} have been deduced in UK and USA from measurements of ^{131}I in forage and milk after the Chernobyl accident (Wilkins *et al.*, 1988; Dreicer & Klusek, 1988). Alderman & Stranks (1967) measured stable iodine in the milk of 18 herds and also in the herbage grazed. There was a good correlation milk/herbage, and, assuming that the cows consumed 12 kg dry weight per day, the transfer factor was $F_m = 2.9 \pm$ (s.e.) 0.3×10^{-3} d l^{-1}.

3.11 ^{131}I in milk following single episodes of deposition

If there is fallout from an accidental emission of fission products, the activity on the ground can be measured quickly by gamma spectrometry of grass and soil. The problem is then to predict the concentrations likely to appear in milk in a few days time. Burton *et al.* (1966) assumed that 25% of the fallout would be intercepted by the edible part of the herbage, taken as amounting to 55 g m^{-2} dry weight. This is equivalent to an absorption coefficient of 4.5 m^2 kg^{-1}, which compares with a value of 2.8 m^2 kg^{-1} obtained in field experiments with ^{131}I vapour (Chamberlain, 1970). Using a value 5×10^{-3} d l^{-1} for F_m, Burton *et al.* calculated the concentration C_I in milk from unit deposition of ^{131}I, obtaining results equivalent to curve A in Fig. 3.6. The peak normalised concentration in milk reached 2 to 3 d after the fallout is $C_I(max) = 0.14$ m^2 l^{-1}.

From day 6 onwards, the slope of the curve corresponds to the effective half-life of ^{131}I on herbage, namely 5 d. Assuming that this continues indefinitely, the area under curve A in Fig. 3.6 is 1.4 m^2 d l^{-1}. This is equivalent to the transfer factor k_m, defined by equation (2.12). Values of F_m for ^{131}I and ^{137}Cs are about the same, but the radioactive decay of ^{131}I reduces k_m compared with that for [137]Cs (Table 2.19). Also shown in Fig. 3.6 are values of C_I as deduced from measurements near

Table 3.6. *Transfer factor of iodine from feed to milk of cattle*

Investigation	F_m (d l^{-1})	Reference
Analysis of feeding experiments:		
UK conditions	5×10^{-3}	Garner (1971)
US conditions	1×10^{-2}	Garner (1971)
Monitoring near US power stations	7×10^{-3} to 3.5×10^{-2}	Hoffman (1978)
Monitoring of [129]I near Sellafield	2×10^{-3}	Wilkins *et al.* (1988)
Measurement of ^{131}I from Chernobyl	2×10^{-3} to 3×10^{-3}	Wilkins *et al.* (1988)
Measurement of ^{131}I from Chernobyl	1×10^{-3}	Dreicer & Klusek (1988)
Measurement of stable I in feed and milk at 18 Welsh farms	3×10^{-3}	Alderman & Stranks (1967)

Seascale, Cumbria and Leeds after the Windscale accident (Booker, 1958; Burch, 1959) and near Seascale and Harwell after Chernobyl (Fulker, 1987; Cambray *et al.*, 1987). The values are lower than the theoretical curve, especially those derived from the Chernobyl fallout, which mostly occurred in heavy rain. As another example, in the Karlsruhe district of Germany, the fallout of ^{131}I in the first few days of May 1986 was 10 kBq m^{-2}. The peak activity in the milk of cows feeding outdoors was 47 Bq l^{-1} (Doerfel & Piesch, 1987), giving a normalised $C_I(max)$ of only 5×10^{-3} m^2 l^{-1}.

Figure 3.7 shows the activity of ^{131}I in thyroids of adult residents of Leeds after the Windscale accident and residents of London after the Chernobyl accident. The dosages of ^{131}I in air were 14 Bqd m^{-3} at Leeds in Oct. 1957 (Stewart *et al.*, 1961) and 4 Bqd m^{-3} in Southern England in May 1986 (Cambray *et al.*, 1987). The points at day 0 in Fig. 3.7 are the calculated activities assuming a breathing rate of 20 m^3 d^{-1}, 60% retention of ^{131}I in the lung, and 30% transfer from lung to thyroid (Medical Research Council, 1975). The remaining points in Fig. 3.7 are the average thyroid activities, as measured by Burch (1959, two sub-

Fig. 3.6. Normalised concentration of ^{131}I in milk after fallout. A, theoretical (Burton *et al.*, 1966); B,C, Seascale and Leeds milk after Windscale accident; D,E, Seascale and Berkshire milk after Chernobyl accident.

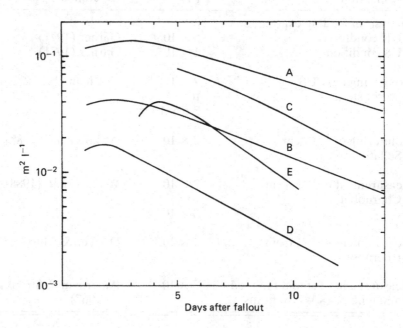

jects) and Hill *et al.* (1986, six subjects). The 1957 Leeds subjects showed increases in thyroid activity, by about a factor 10, as activity ingested, mainly in milk, added to that inhaled. The 1986 London subjects showed a more modest increase. Possible reasons are:

(a) In early May 1986, most dairy cattle were receiving supplementary stored food.

(b) About a third of the Chernobyl ^{131}I was organic iodine, well absorbed in the lung, but not deposited on grass.

3.12 Dose to human thyroids from inhalation and from intake in milk

Table 3.7 shows the dose to the thyroid from inhaling ^{131}I and Table 3.8 shows the dose from ^{131}I in milk (Kendall *et al.*, 1987; Linsley *et al.*, 1986). As usual in hazard evaluation, the calculations are conservative, the assumed intakes and transfers being towards the upper end of the range of possible values. In Table 3.8, the infant thyroid dose per Bq m^{-2} of ^{131}I fallout is derived as follows. The

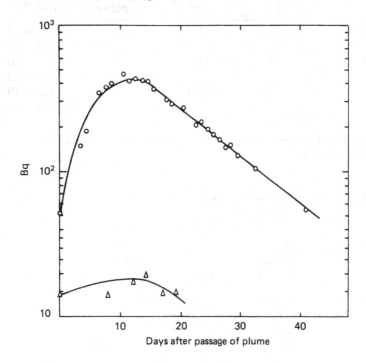

Fig. 3.7. [131]I in thyroids of Leeds residents after Windscale accident (○), and in London residents after Chernobyl accident (△). Points at day zero are calculated from dosage in air.

theoretical transfer factor fallout/milk is 1.4 m^2 d l^{-1} (Section 3.9), so fallout of 1 Bq m^{-2} gives 1.4 Bqd l^{-1} in milk. With milk consumption of 0.71 l d^{-1}, the total intake of ^{131}I is 1.0 Bq. The thyroid dose is 3.7 μSv Bq^{-1}, so the dose per Bq m^{-2} is 3.7 μSv. In the recommendations of the U.S. Food & Drug Administration (1982), a fallout of 130 nCi m^{-2} of ^{131}I is calculated to give a dose of 1.5 rem to the infant thyroid. This is equivalent to 3.1 μSv per Bq m^{-2}.

If v_g is taken as 3 mm s^{-1}, then a dosage of 1 Bqs m^{-3} of ^{131}I in air gives 3×10^{-3} Bq m^{-2} fallout. The ratio of thyroid doses from milk compared with inhalation are then as shown in Table 3.9. The calculated ratio of 20 for adults compares with an increase of about 10 in the thyroid activity of Leeds residents attributed to consumption of milk in the days after the passage of the plume of activity from the Windscale accident (Fig. 3.7). The transfer factor fallout/milk in the Leeds area was lower than that assumed in the calculations leading to Table 3.8

Table 3.7. *Dose to thyroid from ^{131}I in air*

	Thyroid mass (g)	Air intake (m^3 d^{-1})	Thyroid dose (μSv)	
			per Bq intake	per Bqs m^{-3} in air*
Adult	20	23	0.27	7.2×10^{-5} (1.0×10^{-4})
Child (10 a)	7	15	0.74	1.3×10^{-4} (1.4×10^{-4})
Infant (1 a)	1.8	3.8	2.3	1.0×10^{-4} (1.3×10^{-4})

*Figures in parentheses are derived from US Nuclear Regulatory Commission (1977).

Table 3.8. *Dose to thyroid from ^{131}I in milk*

	Milk intake (l d^{-1})	Thyroid dose, μSv	
		per Bq intake	per Bq m^{-2} fallout
Adult	0.82	0.44	0.5
Child	0.92	1.2	1.4
Infant	0.71	3.7	3.7

(compare curves C and A in Fig. 3.6), and the daily consumption of milk was also lower than assumed in Table 3.8.

After the Windscale and Chernobyl accidents, thyroid doses were calculated from measured thyroid activity in children and adults, assuming the mass of the thyroid to be appropriate to the age of the subject. In Table 3.10 the results for children are shown, and converted to the normalised dose per Bq m^{-2} deposition of ^{131}I. Of the cases in Table 3.10, that at Leeds is the only one where the thyroid dose, normalised to unit deposition, approximates to the theoretical calculation of Table 3.8.

In Cumbria, after the Windscale accident, a ban on the sale of local milk was made effective within 3 d. If v_g for ^{131}I in the area of maximum fallout downwind of Windscale was 3 mm s^{-1}, then the measured deposition of 1×10^6 Bq m^{-2} corresponds to an air dosage of 3×10^8 Bqs m^{-3}. From Table 3.7, this would give a theoretical thyroid dose to a child of 40 mSv which corresponds to the lower end of the range given in the first row of Table 3.10. If there had been no ban, and if the transfer from fallout to milk had followed the theoretical curve A in Fig. 3.6, then from Table 3.8 a child drinking 0.92 l of milk per day would have received a thyroid dose of 1.4 Sv.

3.13 Relation between fallout of ^{131}I and external gamma dose rate

If there is an emission of activity, the quickest method of assessing the pattern of fallout is by gamma survey, using instruments carried in vehicles or in aircraft.

To estimate the activity of ^{131}I from the gamma dose requires knowledge of the relative activities of the other fission products. It can be assumed that the other isotopes of iodine (Fig. 3.1) will be present,

Table 3.9. *Ratios of thyroid doses**

	Ingestion/inhalation
Adult	20
Child	32
Infant	110

* Values stated are on the assumption that $v_g = 3$ mm s^{-1}.

Table 3.10. *Dose to children's thyroids from deposition of ^{131}I*

Date	Locality	Fallout of ^{131}I (A_I) (Bq m^{-2})	Children		Dose, D_T (mSv)		D_T/A_I (μSv Bq^{-1} m^2)	Reference
			No.	Age	Median	Range		
1957	Nr Windscale	7×10^5 to 1×10^6	15	n.s.	100	40–160	0.1	Loutit et al. (1960) Chamberlain (1959)
1957	Leeds	8×10^3	1	4	8		1	Burch (1959)
1986	Munich	1.2×10^5	1	1	12*		0.1	Winkelmann et al. (1987)
1986	SE England	1.1×10^3	5	2–12	0.2	0.05–0.5	0.2	Hill et al. (1986) Cambray et al. (1987)

n.s. Not stated.
* Calculated dose from measured activity in diet.

with activities relative to ^{131}I determined by the duration of the irradiation before the release. The activities of isotopes of other elements will depend also on the circumstances of the release.

Figure 3.8 shows the calculated gamma dose rate at 1 m above ground corresponding to a deposition of 1 Bq m[-2] of ^{131}I, as measured at zero time. The fallout is assumed to include:

(A)   All fission products in the proportions of instantaneous fission, as from a nuclear explosion or criticality excursion.

(B)   All fission products in the proportions found in a reactor after 1 a irradiation.

(C)   Volatile fission products (Te, I, Cs) in the same proportion as in (B).

(D)   [131]I only.

The calculations, given in different units by Chamberlain *et al.*

Fig. 3.8. Gamma dose rate from deposited fission products, normalised to initial deposit of 1 Bq m[-2] of [131]I. A, Instantaneous fission products; B, reactor fission products; C, volatile reactor fission products; D, ^{131}I only; E, as measured at Munich after Chernobyl accident.

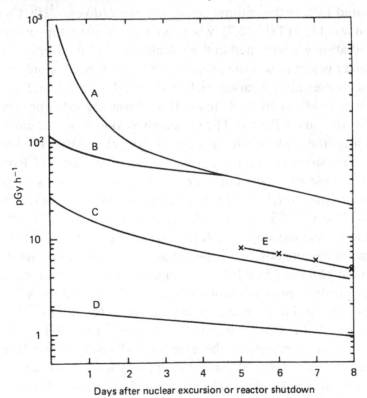

(1961), assume uniform fallout on an infinite plane surface, with the same velocity of deposition for all nuclides, and with no allowance for shielding by surface irregularities. The zero on the time scale in Fig. 3.8 is the time of fission (curve A) or the time of shut down of the reactor (curves B, C). The time of shut down is not necessarily the time of the accident. The Windscale reactor was effectively shut down 4 d before the accident.

Also shown in Fig. 3.8 (curve E) are measurements of gamma dose rate at 1 m above ground on the site of the Institute for Radiation Hygiene at Munich (Winkelmann *et al.*, 1987). Soil cores from the same site showed 5.4 kBq m^{-2} of ^{131}I on 1 June 1986, equivalent to 120 kBq m^{-2} worked back to 26 April, the date of the Chernobyl accident. The activity reached Munich on 30 April, and was deposited in heavy rain. The dose rate recorded on 1 May was 0.92 μGy h^{-1} above background, giving a normalised dose rate of 7.7 pGy h^{-1} per Bq m^{-2} of ^{131}I, at 5 d after the accident. The value calculated for volatile reactor fission products at 5 d (Fig. 3.8, curve C) is 5.8 pGy h^{-1}. Although most of the Chernobyl activity consisted of isotopes of I, Te and Cs, some refractory elements were emitted, and Winkelmann *et al.* calculated that these contributed 13% of the gamma dose. Also the ratio of ^{131}I/^{137}Cs in the emission was 17 : 1 (Table 2.7), whereas a ratio of 20 : 1 appropriate to 1 a of irradiation was assumed in the calculations of Fig. 3.8.

Gamma ray survey was the principal and sometimes the only method of district survey after weapon trials in the 1950s. On 1 March 1954, an H-bomb of 15 MT yield (code name Bravo) was exploded near ground at Bikini Atoll in the Pacific. The radioactive cloud travelled eastwards, and 7 h after the explosion it passed over Rongelap Atoll, 180 km from Bikini. (Glasstone & Dolan, 1977). The 67 inhabitants of Rongelap were evacuated two days later. From subsequent surveys of gamma dose rate on the atoll, it was calculated that the inhabitants received an external dose of 1.75 Gy over the period from 7 to 51 h after the explosion. Using the well established empirical relation for the decay with time of gamma dose from instantaneous fission, it follows that the dose rate at 24 h was 35 mGy h^{-1}. Assuming there was no fractionation of fission products prior to fallout, curve A of Fig. 3.8 shows a dose rate of 250 pGy h^{-1} at 1 d when the activity includes 1 Bq m^{-2} of ^{131}I. Hence the fallout of ^{131}I at Rongelap was 140 MBq m^{-2} (3.8 mCi m^{-2}).

The immediate effects of the external radiation on the Rongelap inhabitants included nausea and a fall in the lymphocyte count to 50% of normal in adults and 25% of normal in children. Long-term follow-up studies (Conrad, 1980) showed an excess of thyroid nodules and thyroid carcinomas in the Rongelap subjects, especially in those who were children in 1954, compared with controls. Estimates of internally deposited isotopes were based on assays of urine samples, taken after evacuation. Based on this evidence, the dose to the children's thyroids from isotopes of iodine is believed to have been in the range 8–18 Sv (Conrad, 1980). Using the above estimate of fallout, this is equivalent to 0.06–0.13 μSv per Bq m^{-2} of ^{131}I. Over half the calculated thyroid dose was due to the short-lived isotopes ^{132}I, ^{133}I and ^{135}I, which are present in high concentrations relative to ^{131}I when the source is recent instantaneous fission. Most of the intake was by inhalation and by drinking rainwater. Cow's milk was not available on Rongelap.

During the 1950s, weapons tests carried out at the Nevada Test Site had a total yield of about 1 MT. The activity of ^{131}I released in these tests, though less by a factor 15 than the activity released at Bikini, was about six times greater than that emitted from Chernobyl (Table 2.3). No measurements of ^{131}I in milk or thyroids were made after these tests, but some were made during further tests, of lower yield, in the 1960s. Knapp (1964) attempted to reconstruct the activities of ^{131}I in the earlier tests from the records of gamma dose rate. He used a relation 1 milliroentgen h^{-1} at 1 d → 1μCi m^{-2} of ^{131}I at 1 d.

One milliroentgen is 84 μGy, and 1 μCi is 3.7×10^4 Bq, so the above relation is equivalent to 227 pGy h^{-1} per Bq m^{-2}, close to the result of Fig. 3.8, curve A. As Knapp emphasised, the relation assumed that there was no chemical fractionation of fission products. If there was fractionation, this might give a higher proportion of volatile fission products, such as iodine, in the long-range fallout (Section 2.3).

Knapp also sought to deduce the maximum concentration $C_I(max)$ of ^{131}I in milk resulting from the fallout, using records of tests in the 1960s where gamma dose and C_I were both measured, Garner's (1960) data on transfer from cattle feed to milk, and Booker's (1958) measurements after the Windscale accident.

On 19 May 1953, a bomb of 32 kT yield (code name Harry) was exploded at the Nevada Test Site. A gamma dose rate equivalent to 27 milliroentgen per hour (230 μGy h^{-1}) at 1 d was measured at St George, Utah, 200 km from the test site. Assuming no chemical fractionation of fission products, the fallout of ^{131}I at St George was 1.0 MBq m^{-2}. This is almost identical to the fallout of ^{131}I in the zone of maximum deposition about 10 km from Windscale (Chamberlain, 1959). Relative

Table 3.11. ^{131}I in milk from weapons tests and Windscale accident

Date	Source	Location	Distance (km)	Release (TBq)	^{131}I peak in milk (Bq l^{-1})	Ratio milk/release (l^{-1})
May 1953	Nevada	St George Utah	200	1.3×10^5	0.3 to 1×10^5 (a)	0.2 to 0.7×10^{-12}
Oct 1957	Windscale	Leeds	150	7×10^2	6.5×10^2	1×10^{-12}

(a) as calculated by Knapp (1964).

to ^{131}I, there would have been more short-lived fission products, and less ^{137}Cs at St George, compared with the Windscale area.

Knapp calculated that $C_I(max)$ would have been in the range 0.7–2.6 μCi l^{-1}(0.026–0.10 MBq l^{-1}) at St George, a range of values which is consistent with measurements near Windscale (Dunster *et al.*, 1958; Booker, 1958). Knapp calculated further that an infant drinking 1 l of milk per day (perhaps more than is usual) would have received a thyroid dose of 120–440 rad (1.2–4.4 Gy). Since the iodine isotopes are β, γ emitters, the dose equivalent in Sv is the same as the dose in Gy.

Knapp's calculations showed the possibility of a public health hazard from the Nevada tests, which, in retrospect, would have justified a ban on the sale of local milk, similar to that applied in Cumbria after the Windscale accident. The calculations were criticised as speculative. However, a further comparison can be made of the consequences of the Harry test and the Windscale accident. St George is 200 km from the Nevada Test Site. Leeds is 150 km from Windscale. In Table 3.11, the calculated $C_I(max)$ levels at Leeds (Burch, 1959) are compared with the respective emissions of ^{131}I.

The Harry test released about 140 times as much ^{131}I as the Windscale accident, and Knapp's calculated $C_I(max)$ at St George is 50–100 times greater than the measured $C_I(max)$ at Leeds. Burch estimated that the dose to the thyroid of a four-year-old Leeds boy, drinking 0.7 l per day of local milk was about 0.8 rad. An infant consuming a similar amount of milk would receive a higher dose because of its smaller thyroid mass. Thus Knapp's estimate of thyroid dose at St George is not unreasonable.

The International Commission on Radiological Protection (1977) introduced the concept of Effective Dose, which takes into account the probability of mortality from cancers in various tissues receiving irradiation. For irradiation of the thyroid, a weighting factor, w_T, equal to 0.03

Table 3.12. *Effective doses at Rongelap and St George, Utah*

	External dose (Sv)	Thyroid dose (Sv)	Thyroid effective dose (Sv)
Rongelap	1.75	8 to 18	0.24 to 0.54
St George	0.03	1.2 to 4.4	0.04 to 0.13

Note. Since the quality factor is unity for external gamma and for irradiation of the thyroid, Sieverts and Grays are equal numerically.

is recommended. This takes into account the slow progress of thyroid cancers and the chances of successful treatment. For external radiation, the weighting factor is unity.

Table 3.12 compares the effective doses from external and thyroid irradiation at Rongelap after the Bikini test and at St George after the Harry test. If the admittedly approximate dose estimates of Conrad *et al.* (1980) are accepted, then the external gamma dose contributed most of the effective dose at Rongelap. The evacuation of the inhabitants after two days reduced the thyroid dose from ingestion of food and water. The doses to the thyroid of the hypothetical infant at St George, Utah, as calculated by Knapp, give effective doses ranging from one to four times the external dose.

The policy of the USAEC in the 1950s of relying on measurements of external dose to monitor fallout from weapons tests appears to be marginally justified in retrospect by the ICRP concept of effective dose.

References

Adams, D.R. & Voillequé, P.G. (1971) Effect of stomatal opening on the transfer of $^{131}I_2$ from air to grass. *Health Physics*, **21**, 771–5.

Alderman, G. & Stranks, M.H. (1967) The iodine content of bulk herd milk in summer in relation to estimated dietary iodine intake of cows. *Journal of Science in Food & Agriculture*, **8**, 151–3.

Amiro, B.D. & Johnston, F.L. (1989) Volatilization of iodine from vegetation. *Atmospheric Environment*, **23**, 533–8.

Aoyama, M., Hirose, K., Suzuki, Y., Inoue, H. & Sugimura, Y. (1986) High level radioactive nuclides in Japan. *Nature*, **321**, 819–20.

Atkins, D.H.F., Chadwick, R.C. & Chamberlain, A.C. (1967) Deposition of radioactive methyl iodide to vegetation. *Health Physics*, **13**, 91.

Barry, P.J. & Chamberlain, A.C. (1963) Deposition of iodine onto plant leaves from the air. *Health Physics*, **9**, 1149–57.

Booker, D.V. (1958) Physical measurement of activity in samples from Windscale. Atomic Energy Research Establishment Report. HP/R 2617. Harwell, Oxon.

Bretthauer, E.W., Mullen, A.L. & Morghissi, A.A. (1972) Milk transfer comparisons of different chemical forms of radioiodine. *Health Physics*, **22**, 257–60.

Bryant, P.M. (1964) Derivation of working limits for continuous release rates of iodine-131 to atmosphere in a milk-producing area. *Health Physics*, **10**, 249–57.

Bunch, D.F. (1968) Controlled environmental radioiodine tests at the National Reactor Testing Station. Report IDO 12063. Idaho Falls, Idaho: NRTS.

Burch, P.R.J. (1959) Measurements at Leeds following the Windscale reactor accident. *Nature*, **182**, 515–19.

Burton, J.D., Garner, R.J. & Scott Russell, R. (1966) Possible relationships between the deposition of fission products and levels of dietary contamination. In *Radioactivity and Human Diet*, ed. R. Scott Russell. Oxford: Pergamon. pp. 459–67.

Cambray, R.S. (1981) Personal communication.

Cambray, R.S. *et al.* (1987) Observations on radioactivity from the Chernobyl accident. *Nuclear Energy*, **26**, 77–110.

Cartan, F.O., Beard, H.R., Duce, F.H. & Keller, J.H. (1968) Evidence for the existence of hypoiodous acid as a volatile iodine species produced in water/air mixtures. Proceedings 10th AEC Air Cleaning Conference. CONF 680821. Springfield VA: NTIS.

Chamberlain, A.C. (1953) Experiments on the deposition of iodine-131 vapour onto surfaces from an airstream. *Philosophical Magazine*, **44**, 1145–53.

(1959) Deposition of iodine-131 in Northern England in October 1957. *Quarterly Journal of the Royal Meteorological Society*, **85**, 350–61.

(1960) Aspects of the deposition of radioactive and other gases and particles. *International Journal of Air Pollution*, **3**, 63–86.

(1981) Emissions of fission products and other activities during the accident to Windscale Pile No 1 in October 1957. Atomic Energy Research Establishment Report M 3194. Harwell, Oxon.

Chamberlain, A.C. & Chadwick, R.C. (1953) Deposition of airborne radioiodine vapour. *Nucleonics*, **11**, 22–5.

(1966) Transport of iodine from atmosphere to ground. *Tellus*, **17**, 226–39.

Chamberlain, A.C., Garner, R.J. & Williams, D. (1961) Environmental monitoring after accidental deposition of radioactivity. *Journal of Nuclear Energy*, **14**, 155–67.

Chameides, W.L. & Davis, D.D. (1980) Iodine: its possible role in tropospheric geochemistry. *Journal of Geophysical Research*, **85**, 7383–98.

Clough, W.S., Cousins, L.B. & Eggleton, A.E.J. (1965) Radioiodine adsorption on particulate matter. *International Journal of Air & Water Pollution*, **9**, 769–89.

Cofman, V. (1919) The active substance in the iodination of phenols. *Journal of the Chemical Society*, **115**, 1040–9.

Collins, R.D. & Eggleton, A.E.J. (1965) Control of gas-borne activity arising from reactor faults. Proceedings Third International Conference on Peaceful Uses of Atomic Energy, vol. 13, pp. 69–74. New York: United Nations.

Conrad, R.A. *et al.* (1980) Review of medical findings in a Marshallese population twenty six years after accidental exposure to radioactive fallout. Report BNL 51261. Upton, N.Y.: Brookhaven National Laboratory.

Crabtree, J. (1959) The travel and diffusion of the radioactive material emitted during the Windscale accident. *Quarterly Journal Royal Meteorological Society*, **85**, 362–70.

Crouch, E.A.C. (1977) Fission product yields from nuclear-induced fission. *Atomic Data & Nuclear Data Tables*, **19**, 417–532.

Devell, L., Aarkrog, A., Blomqvist, L., Magnusson, S. & Tveten, U. (1986) How the fallout from Chernobyl was detected and measured in the Nordic countries. *Nuclear Europe*, **11**, 16–17.

Doerfel, H. & Piesch, E. (1987) Radiological consequences in the Federal Republic of Germany of the Chernobyl reactor accident. *Radiation Protection Dosimetry*, **19**, 223–33.

Dreicer, M. & Klusek, C.S. (1988) Transport of ^{131}I through the grass–cow–milk pathway. *Journal of Environmental Radioactivity*, **7**, 201–7.

Duce, R.A., Zoller, W.H. & Moyers, J.L. (1973) Particulate and gaseous halogens in the Antarctic atmosphere. *Journal of Geophysical Research*, **78**, 7802–11.

Dunster, H.J., Howells, H. & Templeton, W.L. (1958) District surveys following the Windscale accident of October 1957. Proceedings Second International Conference on Peaceful Uses of Atomic Energy, vol. 18, pp. 296–308. Geneva: United Nations.

Eggleton, A.E.J. & Atkins, D.H.F. (1964) The identification of trace quantities of radioactive iodine compounds by gas-chromatographic and effusion methods. *Radiochemica Acta*, **3**, 151–8.

Eisenbud, M. & Wrenn, M.E. (1963) Biological disposition of radioiodine – a review. *Health Physics*, **9**, 1133–9.

Fulker, M.J. (1987) Environmental monitoring by British Nuclear Fuels following the Chernobyl reactor accident. *Journal of Environmental Radioactivity*, **5**, 235–44.

Garland, J.A. (1967) The adsorption of iodine by atmospheric particles. *Journal of Nuclear Energy*, **21**, 687–700.

Garland, J.A. & Cox, L.C. (1984) The uptake of elemental iodine vapour by bean leaves. *Atmospheric Environment*, **18**, 199–204.

Garland, J.A. & Curtis, H. (1981) Emission of iodine from the sea surface in the presence of ozone. *Journal of Geophysical Research*, **86**, 3183–6.

Garland, J.A., Wells, A.C., Higham, E.J. & Smith, I.C. (1984) Experimental study of the deposition of iodine and other fission products in the coolant circuit of a CAGR. In: *Fifth International Meeting on Thermal Nuclear Reactor Safety*, compiled G. Bork & H. Rininsland, vol. 3, pp. 1423–41. Karlsruhe: Nuclear Research Center.

Garner, R.J. (1960) An assessment of the quantities of fission products likely to be found in milk in the event of aerial contamination of agricultural land. *Nature*, **186**, 1063–4.

(1971) Transfer of radioactive materials from the terrestrial environment to animals and man. *CRC Critical Reviews in Environmental Control*, **2**, 337–85.

Glasstone, S. & Dolan, P.J. (1977) The effects of nuclear weapons, 3rd edn. U.S. Depts of Defence and Energy.

Guenot, J., Caput, C., Belot, Y., Bourdeau, F. & Angeletti, L. (1982) Dépôt de l'iode sur les végétaux: influence de l'ouverture stomatique et de l'humidité rélative. Proc. Joint Protection Meeting on Radiological Impact of Nuclear Power Plants on Man and the Environment. Lausanne.

Haller, W.A. & Perkins, R.W. (1967) Organic iodine-131 compounds released from a nuclear fuel processing plant. *Health Physics*, **13**, 733–8.

Heinemann, K. & Vogt, K.J. (1980) Measurements of the deposition of iodine onto vegetation and of the biological half-life of iodine on vegetation. *Health Physics*, **39**, 463–74.

Hill, C.R., Adam, I., Anderson, W., Ott, R.J. & Sowby, F.D. (1986) Iodine-131 in human thyroids in Britain following Chernobyl. *Nature*, **321**, 655–6.

Hoffman, F.O. (1978) A review of measured values of the milk transfer coefficient for iodine. *Health Physics*, **35**, 413–16.

Holland, J.Z. (1963) Physical origin and dispersion of radioiodine. *Health Physics*, **9**, 1095–103.

Hood, E.M. & Clough, P.N. (1984) The behaviour of iodine in Advanced Gas-cooled Reactor (AGR) circuits. In: *Fifth International Meeting on Thermal Nuclear Reactor Safety*, compiled G. Book & H. Rininsland. vol. 3, pp. 1432–41. Karlsruhe: Nuclear Research Center.

International Commission on Radiological Protection (1977) *Recommendations of the ICRP*, ICRP Publication 26. Oxford, Pergamon.

Islitzer, N.F. (1962) The role of meteorology following the nuclear accident in south east Idaho. Report IDO 19310. National Reactor Testing Station, Idaho Falls, Idaho.

Jenkin, M.E., Cox, R.A. & Candleland, D.E. (1985) Photochemical aspects of tropospheric iodine behaviour. *Journal of Atmospheric Chemistry*, **2**, 359–75.

Keller, J.H., Duce, F.A., Ponce, D.T. & Maeck, W.J. (1970) Hypoiodous acid: and airborne inorganic iodine species in steam–air mixtures. *Proc. 11th A.E.C. Air Cleaning Conference*, ed. M.W. First & J.M. Morgan, pp. 467–79. CONF 700 816. NTIS, Springfield, VA.

Kendall, G.M., Kennedy, B.W., Greenhalgh, J.R., Adams, N. & Fell, T.P. (1987) Committed doses to selected organs and committed effective doses from intake of radionuclides. Report GS7. National Radiological Protection Board. Chilton, Oxon.

Knapp, H.A. (1964) Iodine-131 in fresh milk and human thyroids following a single deposition of nuclear test fallout. *Nature*, **202**, 534–7.

Leifer, R., Helfer, I., Miller, K. & Silvestri, S. (1986) Concentrations of gaseous ^{131}I in New York City air following the Chernobyl accident. In: Report EML 460 (ed. H.L. Volchok). US Dept of Energy. NTIS, Springfield, VA.

Linsley, G.S., Crick, M.J., Simmonds, J.R. & Haywood, S.M. (1986) Derived emergency reference levels. Report D.L.10. Chilton, Oxon: National Radiological Protection Board.

Loutit, J.F., Marley, W.G. & Russell, R.S. (1960) The nuclear reactor accident at Windscale in October 1957: environmental aspects. In: *The Hazards to Man of Nuclear and Allied Radiations*. Cmnd 1225. London: HMSO.

Lovelock, J.E., Maggs, R.J. & Wade, R.J. (1973) Halogenated hydrocarbons in and over the Atlantic. *Nature*, **241**, 194–6.

Marter, W.L. (1963) Radioiodine release incident at the Savannah River Plant. *Health Physics*, **9**, 1105–9.

Medical Research Council (1975) *Criteria for Controlling Radiation Doses to the Public after Accidental Escape of Radioactive Material*. London: HMSO.

Megaw, W.J. & May, F.G. (1962) The behaviour of iodine released in reactor containments. *Reactor Science & Technology*, **16**, 427–36.

Miyake, Y. & Tsunogai, S. (1963) Evaporation of iodine from the ocean. *Journal of Geophysical Research*, **68**, 3989–93.

Morgan, A., Morgan, D.J. & Arkell, G.M. (1966) A study of the retention and subsequent metabolism of inhaled methyl iodide. In: *Inhaled Particles and Vapours II*, ed. W.H. Walton, pp. 309–21. Oxford: Pergamon.

Morris, J.B., Diffey, H.R., Nicholls, B. & Rumary, C.H. (1962) The removal of low concentrations of iodine from air on a plant scale. *Reactor Science & Technology*, **16**, 437–45.

Moyers, J.L. & Duce, R.A. (1972) Gaseous and particulate iodine in the marine atmosphere. *Journal of Geophysical Research*, **77**, 5229–38.

Moyers, J.L., Zoller, W.H. & Duce, R.A. (1971) Gaseous iodine measurements and their relationship to particulate lead in a polluted atmosphere. *Journal of Atmospheric Sciences*, **28**, 95–8.

Noguchi, H. & Murata, M. (1988) Physicochemical speciation of airborne ^{131}I in Japan from Chernobyl. *Journal of Environmental Radioactivity*, **7**, 65–74.

Parker, H.M. (1956) Environmental radiation exposure. *Proceedings International Conference on Peaceful Uses of Atomic Energy*, vol. 13, pp. 305–10.

Peirson, D.H. & Keane, J.R. (1962) Characteristics of early fallout from the Russian nuclear explosions of 1961. *Nature*, **196**, 801–7.

Perkins, R.W. (1963) Physical and chemical form of ^{131}I in fallout. *Health Physics*, **9**, 1113–22.

Rahn, K.A., Borys, R.D. & Duce, R.A. (1976) Tropospheric halogen gases: inorganic and organic components. *Science*, **192**, 549–50.

Robens, E. & Aumann, D.C. (1988) Iodine-129 in the environment of a nuclear fuel reprocessing plant: I. ^{129}I and ^{127}I contents of soils, food crops and animal products. *Journal of Environmental Radioactivity*, **7**, 159–75.

Robens, E., Hauschild, J. & Aumann, D.E. (1989) Iodine-129 in the environment of a nuclear fuel re-processing plant: IV ^{129}I and ^{127}I in undisturbed surface soils. *Journal of Environmental Radioactivity*, **9**, 17–29.

Soldat, J.K. (1976) Radiation doses from iodine-129 in the environment. *Health Physics*, **30**, 61–70.

Stewart, N.G. & Crooks, R.N. (1958) Long range travel of the radioactive cloud from the accident at Windscale. *Nature*, **182**, 627–30.

Stewart, N.G., Crooks, R.N. & Fisher, E.M.R. (1961) Measurements of the radioactivity of the cloud emitted from Windscale: data submitted to the I.G.Y. Report HP/M 857. Harwell, Oxon: Atomic Energy Research Establishment.

Stewart, S.P. & Wilkins, B.T. (1985) Aerial distribution of ^{129}I in West Cumbrian soils. *Journal of Environmental Radioactivity*, **2**, 175–82.

United Nations Scientific Committee on Effects of Atomic Radiations (1982) Ionising radiation: sources and biological effects. New York, United Nations.

U.S. Food & Drug Administration (1982) Accidental contamination of human foods and animal foods. Report FR 471205, pp. 58798–9.

U.S. Nuclear Regulatory Commission (1977) Calculation of annual doses to man from routine releases of reactor effluents. Regulatory Guide 1.109, Rev.1.

Voillequé, P.G. (1979) Iodine species in reactor effluents and in the environment. Report NP 1269. Electric Power Research Institute. Palo Alto. CA.

Voillequé, P.G. & Keller, J.H. (1981) Air to vegetation transport of ^{131}I as hypoiodous acid (HOI). *Health Physics*, **40**, 91–4.

Wershofen, H. & Aumann, D.C. (1989) Iodine-129 in the environment of a nuclear fuel reprocessing plant. VII Concentrations and chemical forms of ^{129}I and ^{127}I in the atmosphere. *Journal of Environmental Radioactivity*, **10**, 141–56.

Whitehead, D.C. (1984) The distributions and transformations of iodine in the environment. *Environment International*, **10**, 321–39.

Wilkins, B.T., Bradley, E.J. & Fulker, M.J. (1988) The influence of different agricultural practices on the transfer of radionuclides from pasture to milk after the Chernobyl accident. *The Science of the Total Environment*, **68**, 161–72.

Winkelmann, I., *et al.* (1987) *Radioactivity Measurements in the Federal Republic of Germany after the Chernobyl Accident*. Munich: Institut für Strahlenhygiene des Bundesgesundheitamtes, Neuherberg.

4

Tritium

4.1 Definitions

Tritium (^{3}H) has a radioactive half-life of 12.3 a and decays to ^{3}He by emitting a soft beta particle, maximum energy 18 keV. Measurements of tritium are given variously in terms of:

$$\text{Tritium unit (TU)} = \frac{(10^{18} \times \text{no. of T atoms})}{(\text{no. of H atoms}) \text{ in sample}}$$

$$\text{Mixing ratio} = \frac{(\text{mass of T})}{(\text{mass of sample})}$$

$$\text{Activity} = \text{Bq or Curies of T in sample}$$

Useful conversions are:

$$1 \text{ kg T} = 9.7 \text{ MCi} = 3.6 \times 10^{17} \text{ Bq} \qquad (4.1)$$

$$1 \text{ TU in water} = 3.26 \text{ pCi l}^{-1} = 0.12 \text{ Bq l}^{-1} \qquad (4.2)$$

The boiling point of tritiated water (HTO) is 100.76°C. The vapour pressure of HTO is slightly less than that of H_2O, and

$$\beta = (\text{vapour pressure } H_2O)/(\text{vapour pressure HTO})$$

has the value 1.1 at ambient temperatures, reducing to 1.025 at 100°C (Sepall & Mason, 1960; Jacobs, 1968).

Tritium is present naturally in the atmosphere, but the amounts were increased greatly in the late 1950s and 1960s by production and testing of thermonuclear weapons. Tritium is also a fission product and activation product produced in power reactors. Releases occur from reactors and reprocessing plants. Its use will increase greatly if fusion power is developed.

4.2   Natural tritium

Tritium is formed in the upper atmosphere by bombardment of nitrogen, oxygen and carbon by high energy protons and by capture of cosmic-ray generated neutrons by nitrogen

$$^{14}N + n = {}^{12}C + {}^3H$$

The rate of production depends on solar activity and ranges from 0.4 to 0.8 kg (4 to 8 MCi) per annum (National Council on Radiation Protection and Measurement, 1979). Most of the tritium is formed in the stratosphere, and remains there for some months, giving an inventory of 0.24 kg (2.3 MCi) (UNSCEAR, 1977).

Harteck (1954) measured the amount of T in the lower atmosphere before H-bomb tests had added significantly to natural activity, and found 4000 and 3 TU in hydrogen and water vapour respectively. The amount of H_2 in the atmosphere has increased in recent years due to industrial production (Schmidt, 1974). Circa 1950 there were about 1.5×10^{11} kg of H as H_2, compared with 1.4×10^{15} kg as H_2O. Thus Harteck's values for natural T correspond to tropospheric inventories of 1.8 g as HT and 13 g as HTO. Because HTO is deposited in rain and by vapour transfer to the sea, its residence time in the troposphere is only about a week, similar to that of ^{137}Cs (Fig. 2.8), and more than 90% of the atmospheric inventory is in the stratosphere. By contrast, the residence time of HT in the troposphere is several years (Mason & Ostlund, 1979), and most of the atmospheric inventory of HT is in the troposphere. It is only necessary for a small fraction of naturally produced T to form HT to account for the high specific activity of hydrogen gas.

4.3   Tritium from weapons tests and reactors

Begemann & Libby (1957) estimated that 1.1 kg of T was released to atmosphere for each megatonne (MT) thermonuclear explosion. The tests between 1954 and 1963 had a fusion yield of 320 MT. Allowing for radioactive decay, the global inventory in 1963, including tritium in the atmosphere, groundwater and oceans, was about 330 kg. French and Chinese thermonuclear tests between 1968 and 1977 may have added another 20–30 kg. In 1972, by which time most of the pre-1963 tritium had returned to the earth's surface, a world-wide survey of oceanic waters gave a total of 164 kg (Ostlund & Fine, 1979). Corrected for radioactive decay, this is equivalent to an inventory of 270 kg in 1963.

Most of the T from atmospheric tests is formed as HTO. Presumably, the heat of the fireball ensures oxidation. Ehhalt (1966) noted an increase in the atmospheric HT before the series of large thermonuclear tests in 1962 and 1963, but observed no increases correlated with individual tests. Mason & Ostlund (1979) analysed samples of stratospheric air after the Chinese tests of November 1976 and observed a big increase in HTO but not in HT. HT may, however, be released following underground tests.

Lithium is irradiated in reactors to produce tritium using the (nα) reaction in lithium 6. During processing of the product, tritium may be released as HT, HTO or CH_3T. Murphy & Prendergast (1979) estimated that annual releases of T from the Savannah River Plant averaged 80 g, with 40% as HT and 60% as HTO. After the Windscale accident, a peak in the tritium content of atmospheric hydrogen was observed at Hamburg (Ehhalt & Bainbridge, 1966). It was calculated that about 10 g of HT were released from the reactor (Chamberlain, 1981). Leakages from production plants are the main source of reduced forms of tritium in the atmosphere.

Tritium is also formed as a product of ternary fission in power reactors. Yields in thermal fission of ^{235}U and ^{239}Pu are about 1×10^{-4} and 1.5×10^{-4} respectively (NCRP, 1979). Most of the tritium is retained in the fuel, but some may be released to atmosphere as HTO during reprocessing. At present, no fuel is reprocessed in the USA. The NCRP (1979) report included speculative estimates that reprocessing in Europe, excluding the USSR, may release 0.4 kg a^{-1}. This is small compared with the release in atmospheric thermonuclear tests.

4.4   Transport of HTO in the atmosphere

In common with other nuclides of stratospheric origin, downwards transport of HTO into the troposphere and thence to the surface is most rapid in spring. Figure 4.1 shows the seasonal variation of HTO and ^{90}Sr in continental air (Ehhalt, 1971). Removal of HTO from the atmosphere is by incorporation into raindrops and by diffusive gaseous exchange at the earth's surface. About 70% of the HTO falling in rain on land re-evaporates within a few weeks. The remainder runs off to sea. Water vapour in air near the ground, rain, and surface water all contain similar concentrations of HTO. Re-evaporation from land is a secondary source of HTO vapour, returning precipitated activity to the atmosphere, and this may account for the lag in the peak of T compared with ^{90}Sr in Fig. 4.1. By contrast, the oceans are a sink, since the

156   Tritium

residence time of water in the sea is long compared with the radioactive half-life of tritium.

Figure 4.2 shows the concentration of HTO in rainfall, in TU units, at Valentia on the southwest coast of Ireland from 1952 to 1972 (Weiss *et al.*, 1979). Measurements at sea showed no difference in TU of rain and water vapour, when allowance was made for the isotopic difference in vapour pressure of HTO and H_2O. TU levels in water vapour and rain are higher well inland than at the coast by factors up to four (Eriksson, 1965). There are two reasons for this difference. The rainfall and humidity are usually less in continental areas, so that the downwards flux of HTO is less diluted with H_2O. Also, re-evaporation of HTO from land increases the atmospheric concentration at low levels.

The pre-bomb HTO concentration in rain at Valentia, on the S.W. coast of Ireland, was 2.5 TU. In European and American wine (as measured later) it was 3 to 5 TU (Schell *et al.*, 1974). By 1963, the concentration in rain at Valentia had reached 1,000 TU. It then declined rapidly, though Chinese and French tests reduced the fall after 1967. The reduction from 1963 to 1967 was by a factor of 12. Allowing for

Fig. 4.1. Seasonal variation of tritium ($\triangle$) and ^{90}Sr ($\bigcirc$) in air near the ground. (After Ehhalt, 1971.)

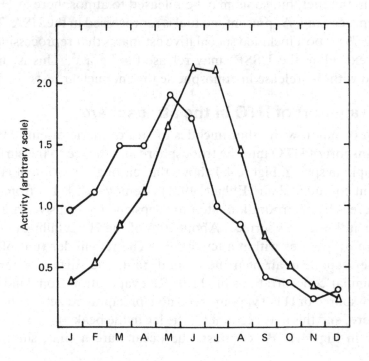

radioactive decay, this corresponds to a stratospheric residence time of 15 months.

Transports of HTO vapour and H_2O vapour to and from surfaces are controlled similarly by eddy diffusion in the free air and molecular diffusion across the viscous boundary layer near the surface. There is also a liquid phase boundary layer, and diffusion through this is a limiting resistance to the transport of sparingly soluble gases such as H_2 or HT. For HTO, the liquid film resistance is negligible (Slinn *et al.*, 1978). When the concentration gradients are in opposite directions, transport of HTO to a water surface can proceed simultaneously with evaporation of H_2O.

If χ_1, χ_0 are the vapour densities ($kg\,m^{-3}$) of HTO in the free air and at the surface, q_1, q_0 the vapour densities of H_2O, and A_T the mass ratio HTO/H_2O in the water (not the same as the TU, which is the atom ratio) then

$$\chi_0 = \frac{A_T q_0}{\beta} \tag{4.3}$$

Fig. 4.2. Tritium in rain at Valentia (L.-H. scale) and tritium in hydrogen at various locations (R.-H. scale).

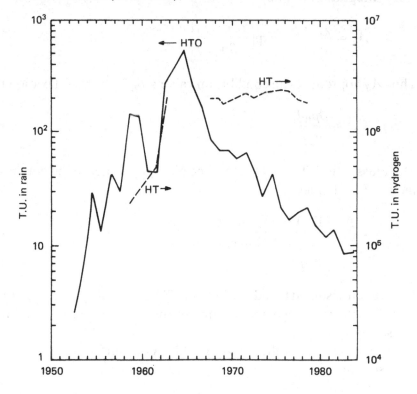

The factor β, which is about 1.1 (Sepall & Mason, 1960), allows for the difference in vapour pressure of HTO compared with H_2O.

The flux ($kg\ m^{-2}\ s^{-1}$) of HTO to the surface is

$$Q = v_g(\chi_1 - \chi_0) \tag{4.4}$$

where v_g is the velocity of deposition, transfer velocity, or conductance. The reciprocal of v_g is the resistance, r. In the electrical analogy, Q corresponds to the current and $\chi_1 - \chi_0$ to the E.M.F. Either the conductance or the resistances in diffusion pathways may be additive, according to the circumstances.

If L is a characteristic length defined as the ratio of the volume of a body of water to its exposed area, and loss of water by evaporation is neglected, the rate of change of the HTO/H_2O mixing ratio in the liquid phase due to deposition of HTO vapour is given by

$$\frac{dA_T}{dt} = \frac{Q}{\rho_W L} = \frac{v_g(\chi_1 - A_T q_0 \beta^{-1})}{\rho_W L} \tag{4.5}$$

where ρ_W is the density of water (1,000 in SI units).

By integration of (4.5)

$$A_T = \frac{\chi_1 \beta}{q_0}\left[1 - \exp\left(-\frac{v_g q_0 t}{\beta \rho_W L}\right)\right] \tag{4.6}$$

Thus A_T approaches an equilibrium value $\chi_1 \beta q_0^{-1}$ with a relaxation time

$$\lambda_A^{-1} = \frac{\beta \rho_W L}{v_g q_0} \tag{4.7}$$

The mass ratio HTO/H_2O in the free air is $\chi_1 q^{-1}$, hence the equilibrium factor, EF, defined

$$EF = \frac{\text{HTO/}H_2O \text{ in liquid}}{\text{HTO/}H_2O \text{ in gas}} \tag{4.8}$$

$$= \beta q_1 q_0^{-1} \tag{4.9}$$

It has been assumed that diffusion of HTO is the only process operating, and that there is negligible change in the water content of the liquid phase.

Three examples of the transport of HTO to surfaces are given in the next section.

4.5   Deposition of HTO

(a) Ocean surface

The effective depth of the ocean is so great that the HTO concentration in seawater never approaches the equilibrium value relative to the concentration in air. So the sea is a perfect sink for HTO vapour and χ_0 is effectively zero. The transfer velocity, v_g, of HTO from atmosphere to sea is nearly the same as the transfer velocity of H_2O vapour during evaporation from the sea (the difference in molecular diffusivity as between HTO and H_2O is not important in this connection). Measurements in various oceans (Brutsaert, 1982) have given a value about 1.3×10^{-3} for the bulk transfer coefficient defined

$$Ce_{10} = \frac{E}{(q_0 - q_{10})\, u_{10}} = \frac{v_g}{u_{10}} \qquad (4.10)$$

where E is the evaporative flux of water vapour, u the wind speed, and the suffix 10 refers to a height of 10 m above the sea surface. There is little dependence of Ce_{10} on u_{10}, so v_g is proportional to u_{10}, and, if this is taken as 8 m s^{-1}, v_g is about 10 mm s^{-1}. If winds flow over the sea for several days, depletion of HTO in the lower atmosphere up to a height of several thousand metres can be expected.

(b) Raindrops

Equation (4.7) can be used to calculate the relaxation time for incorporation of HTO into raindrops (Chamberlain & Eggleton, 1964). The transport velocity to a spherical drop is given by Frössling's equation:

$$\text{Sh} = 2\,(1 + 0.276\,\text{Sc}^{1/3}\,\text{Re}^{1/2}) \qquad (4.11)$$

where: Sh = Sherwood number = $v_g d_p D^{-1}$, d_p = diameter of drop, D = molecular diffusivity of water vapour (2.3×10^{-5} m^2 s^{-1} at 20°C), Sc = Schmidt number = u/D, v = kinematic viscosity of air (1.5×10^{-5} m^2 s^{-1} at 20°C), Re = Reynolds number = $v_s d_p v^{-1}$, and v_s = velocity of drop relative to air.

Equation (4.11) applies provided that the gas phase resistance to the transfer of HTO is limiting. This requires that the accommodation coefficient of molecules at the surface, the solubility, and the liquid phase diffusion, or mixing, within the drop, are all sufficiently high.

A raindrop of diameter 1 mm has a fall velocity $v_s = 4.0$ m s^{-1}, giving Re = 270, Sh = 9.6, $v_g = 0.22$ m s^{-1}. For a sphere, $L = d_p/6$, and if q_0 is taken as 0.018 kg m^{-3}, corresponding to saturation vapour pressure at

20°C, the relaxation time λ_A^{-1} derived from (4.7) is 46 s. Figure 4.3 shows λ_A^{-1} plotted against the diameter of the raindrop. Friedman *et al.* (1962) and Booker (1965) measured the relaxation time experimentally by allowing water drops to fall through HTO vapour in saturated air, and their results are also shown in Fig. 4.3. From (4.7), λ_A^{-1} is inversely proportional to the saturation water vapour density q_0, so the relaxation time is longer at lower temperatures. The distance the drop falls in the relaxation time is $\lambda_A^{-1} v_s$, and from Fig. 4.3 this distance is 180 m for a 1-mm drop. Thus the HTO/H$_2$O mixing ratio in a raindrop rapidly equilibrates with the mixing ratio in air.

The annual rainfall over the oceans in middle latitudes is about 1 m,

Fig. 4.3. Relaxation time of absorption of HTO vapour by raindrops. Experimental results of Booker, 1965 (○) and Friedman *et al.*, 1962 (●). Line is theoretical, derived from equations (4.7) and (4.11).

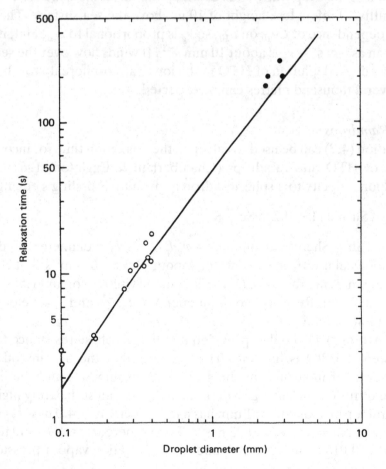

equivalent to 30 mg m^{-2} s^{-1}. If the H$_2$O vapour density is 12 g m^{-3} (75% R.H. at 17.5°C) the ratio of the flux of H$_2$O to the concentration in air (that is the velocity of deposition in rain) is 2.5 mm s^{-1}. If HTO in raindrops has reached equilibrium with HTO in air, v_g for deposition of HTO in rain is also 2.5 mm s^{-1}. This is only about one quarter of v_g for vapour transport of HTO to the sea, as derived above. So diffusive transport to the oceans is the most important mechanism of removal of HTO from the atmosphere, as originally pointed out by Eriksson (1965). This process depletes HTO in water vapour near the sea surface, and, because HTO exchanges between vapour and raindrops, there is less HTO in oceanic than in continental rain. Vapour transport of HTO to the ground is much more important than washout in rain near local sources of emission, because in these circumstances the vertical extent of the plume of HTO vapour is less than the relaxation length for absorption by raindrops (Chamberlain & Eggleton, 1964). Amano & Kasai (1988) found an average of 1.7 Bq l^{-1} (14 TU) in rain near the Tokai Research Establishment in 1984. This is not much more than levels elsewhere in the Northern Hemisphere at that time (Fig. 4.2). HTO in air at Tokai ranged from 10 to 100 Bq per litre of moisture, due to discharge from reactors and treatment plants, and due also to re-evaporation from the soil of previously-deposited HTO.

(c) Vegetation and soil

The diffusion path between the atmosphere and the mesophyll of leaves is through pores, termed stomata, which are under physiological control, shutting at night and when the plant is under water stress. Transport of water vapour from the leaf is controlled by the resistance (r_s) through the stomata and the resistance (r_a) of the boundary layer over the leaf. The total leaf resistance ($r_l = r_a + r_s$) of herbaceous plants is typically 20 to 200 m^{-1} s when the stomata are open (Rutter, 1975). If the stomata are closed and the cuticle undamaged, the leaf resistance increases by one or two orders of magnitude. Desert plants are adapted to have high leaf resistances. Most leaves have stomata on one side only and the resistances are calculated relative to the area of one side of the leaf.

Belot *et al.* (1979) exposed vines (*Vitis vinifera*) in the field to HTO vapour. Garland & Cox (1982) grew bean plants (*Vicia faba*) in compost and exposed their leaves in a wind tunnel. Both sets of authors estimated the mass of water per unit area of leaf (the parameter $\rho_W L$ in equation 4.7). Belot *et al.* estimated the leaf resistance to water vapour by porometry and Garland & Cox by measuring the rate of loss of water from the plants.

Figure 4.4. shows the ratio (A_T/χ_1) of the HTO/H_2O mixing ratio in the leaf water to the concentration of HTO in the air, in one series of Belot *et al.*'s experiments and one series of Garland & Cox's. As it happened, the values of the parameters in equation (4.7) in these experiments were similar (Belot *et al.*, $r_l = 40$ s m^{-1}, $L = 1.6$ mm, $q_0 = 18.4 \times 10^{-3}$ kg m^{-3}; Garland & Cox, $r_l = 43$ s m^{-1}, $L = 2.0$ mm, $q_0 = 17.4 \times 10^{-3}$ kg m^{-3}). The values of A_T/χ_1 from equation (4.7) with these two sets of constants (replacing v_g by $1/r_e$) are shown in Fig. 4.4. The relaxation times λ_A^{-1} from equation (4.7) are 1.1 h and 1.5 h respectively.

Kline & Stewart (1974) had previously found a mean relaxation time of 1.6 h for uptake of HTO by grass leaves in daylight. Stems were labelled much more slowly, with λ_A^{-1} about 50 h. Any diffusion downwards from leaves to stems would be against the transpiration flow, and the labelling was probably via uptake of HTO to soil.

The equilibrium factor, *EF* (ratio of TU in plant to TU in atmospheric water vapour), is $\beta q_1 q_0^{-1}$ (equation (4.9)) and is usually less than unity.

Fig. 4.4. Uptake of HTO by leaves from atmosphere: Belot *et al.* (1979), experimental ×, theory ———; Garland & Ameen (1979), experimental ○, theory – – – –.

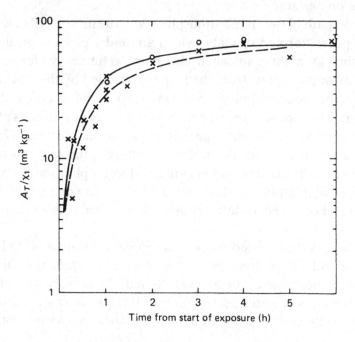

In daytime, leaf temperatures exceed air temperatures and relative humidity in the free air is less than 100%, so $q_1 < q_0$. For example if leaf temperature is 20°C, air temperature 15°C and R.H. 75%, $q_1/q_0 = 0.54$ and $EF = 0.6$. The plants maintain saturation in the mesophyll by absorbing water through their roots, and this water has a relatively low HTO concentration unless the atmospheric HTO has been maintained long enough to bring groundwater into equilibrium. When maize and barley were grown for 30 d from seed in a sealed enclosure, the HTO concentration in the plant water was 90% of that of the atmospheric water (Garland & Ameen, 1979). In the vicinity of the Tokai nuclear facilities where emissions of HTO vapour had continued for some years, Amano & Kasai (1988) found equilibrium factors averaging 0.76 and 0.88 for soil/air and vegetation/air, respectively.

The interaction of atmospheric HTO with vegetation and soil has several time scales. After a short release, airborne HTO is deposited to land or sea by vapour diffusion with a v_g of about 10 mm s^{-1} (Garland, 1980). From land, about 75% of the HTO returns to the atmosphere within days or weeks depending on climate. At sea, HTO mixes downwards and only a small fraction re-enters the atmosphere within the radioactive life of tritium.

When atmospheric HTO remains at a more or less constant level for months, water in soil and surface run-off approaches equilibrium with the atmosphere. In 1963, tritium in US surface water averaged 3000 pCi l^{-1} (110 Bq l^{-1}) (NCRP, 1979), equivalent to 920 TU and rather greater than the level in oceanic rains (Fig. 4.2).

4.6   Uptake of HTO by human body

Ingestion, inhalation and passage through the skin all contribute to the exchange of water and HTO between the human body and the environment. The volumes of water that can be exchanged by these routes are shown in Table 4.1 (NCRP, 1979). It is assumed that all HTO in the inhaled air will exchange in the lung. Osborne (1966) exposed volunteers to HTO vapour in a chamber, and found that uptake by vapour exchange through the skin was equivalent to the absorption from 10 l of air per min (1.6×10^{-4} m^3 s^{-1}). This compares with 16 l per min exchanged by inhalation during moderate activity. The area of skin averaged 1.9 m^2, so the skin resistance to vapour transfer was 1.2×10^4 s m^{-1}, about the same as the resistance of leaf cuticle when the stomata are closed (Rutter, 1975). The saturation H$_2$O vapour density at body temperature is 47 g m^{-3}, giving a driving force of about 40 g m^{-3} relative to ambient air. Hence the loss of water in insensible sweat should be 400 mg min^{-1} if the outwards flux is equivalent to the excess vapour in 10 l min^{-1}. The 'standard man's' loss of water in sweat is 300 g d^{-1}, or 208 mg min^{-1}.

Table 4.1. *Volumes of water available for daily exchange with the human body*

Source	l d^{-1}
Drinking water	1.22
Food, milk, other fluids	1.56[a]
Inhalation	0.13[b]
Passage through skin	0.09[c]
Total	**3.0**

(a) Including 0.25 l d^{-1} by oxidation of food.
(b) Assuming 22 m^3 d^{-1} inhaled with vapour density 6 g m^{-3}.
(c) As estimated by Osborne (1972).

The total exchange of 3 l d^{-1} water volume from Table 4.1, when related to an adult water content of 41 l, gives a relaxation time of 14 d (half-life 10 d) for exchange of HTO with body water. Numerous measurements of retention half-time in individuals accidentally contaminated with HTO have given a mean half-life of 9.5 d (range 4–18 d) (Butler & LeRoy, 1965). The half-life depended on the consumption of fluids and was shorter in summer than in winter. In addition to the readily exchanged HTO in the body, some tritium becomes organically bound, and the compounds exhibit biological half-lives ranging from 20 to 500 d (NCRP, 1979). However, the organically bound component is small. For three years, Moghissi *et al.* (1987) reared successive generations of rabbits on alfalfa grown in tritiated water. The T/H ratio in the tissues of the rabbits remained essentially the same as in the water.

The dose to tissue from HTO is low, because the energy of decay is low and the distribution uniform. In the 1960s, the average dose to man in the northern hemisphere from tritium reached a peak of 2 μGy a^{-1}, about 0.2% of natural background radiation.

4.7   Transport and deposition of HT

The tritium content of atmospheric hydrogen is shown in Fig. 4.2.

The results for 1954–68, from Ehhalt (1966) and Friedman & Scholz (1974) were obtained by measuring tritium in the hydrogen fraction of liquid air, and include corrections for the presence of industrial hydrogen, which has a low T/H ratio because it is made from water. In more recent work (Mason & Ostlund, 1979), the tritium was measured by drawing air through a multistage sampler. In the first stage, a molecular sieve removed water and HTO. Then a bed of colloidal platinum oxidised HT to HTO which was removed in a second sieve. Finally a furnace with an oxidising agent was used to decompose organic compounds such as CH_3T and the product was removed in a third sieve.

In Fig. 4.2, Mason & Ostlund's results for 1969–78 have been converted from T atoms per mg air to TU assuming that H_2 in air is 0.038 ppm by mass. The sources of HT, unlike those of HTO, have been mainly in the troposphere. Mason & Ostlund showed that, by 1978, HT had become well mixed in tropospheric air of both hemispheres, with a concentration of 1.7×10^6 TU (40 T atoms per mg air). The global inventory of T as HT was 1 kg, assuming a global inventory of H_2 of 2×10^{11} kg. By comparison, the atmospheric inventory of HTO, which comprised about 300 kg T in 1963, had declined to 5.3 kg in 1978.

By considering the mass balance of HT in the southern hemisphere, where there are no artificial sources other than through inter-hemispheric transfer, Mason & Ostlund estimated that HT is removed from the atmosphere with a rate constant 0.155 a^{-1}, giving a mean residence time of 6 a. The main sink for H_2 or HT is oxidation by soil bacteria (Schmidt, 1974; Garland & Cox, 1980; Sweet & Murphy, 1981). Land comprises 29% of the earth's surface. The effective depth of the atmosphere (mass per unit ground area divided by density at ground level) is 8000 m and HT is well mixed. It follows that a deposition velocity to land of 0.135 mm s^{-1} would give a removal constant of 0.155 a^{-1}.

4.8   Field measurements of deposition of tritium gas

Garland & Cox (1980) placed PVC enclosures over soil at various sites in southern England, introduced T_2 vapour, and measured the rate at which it disappeared. The loss was exponential, and assuming it was caused by deposition to the ground the velocity of deposition averaged 0.4 mm s^{-1}. This is three times the value deduced for HT in the previous section, but it is not known whether there is a real difference between deposition of HT and T_2. There was some seasonal variation in deposition of T_2 which was probably related to soil porosity. Extraction with water showed conversion from T_2 to HTO in the soil, and supplementary experiments showed that the sink was in the soil, not the vegetation. Uptake of T_2 by seawater was measurable, but was less than 1% of uptake by soil of equivalent surface area. In similar experiments in South Carolina, Sweet & Murphy (1981) found a mean v_g to soil of 0.2 mm s^{-1}. Uptake was inhibited by saturating the soil with water or by treating it with ethanol to kill bacteria.

Uptake by soil and conversion of molecular tritium to HTO had previously been observed by Murphy et al. (1977) after two accidental releases from the Savannah River plant. It was estimated that 1.8×10^{16} Bq of tritium were released in May 1974 and 6.7×10^{15} Bq in December 1975. HTO in soil cores taken, on the plume axis 12 km downwind, 8 d after the second release showed a peak at 0.15 m depth, owing to seepage resulting from rain. From the integrated profile, the total deposition was 1.7 MBq m^{-2}. The dosage (airborne concentration multiplied by time), of molecular tritium was estimated at 1300 MBqs m^{-3} so v_g was 1.3 mm s^{-1}, higher than expected from field experiments. After both releases from Savannah River, the HTO concentrations (Bq T per ml H_2O) were higher in soil than in vegetation, confirming that oxidation was in the soil. After the release in May 1974, HTO in herbaceous vegetation peaked at about 1 d, and in pine needles 5 d after the deposition. Subsequently, HTO in pine needles showed a two-phase exponential decline, with half-lives of 2.4 and 23 d.

In 1979/1980, Sweet & Murphy (1984) made a survey of tritium in pine needle litter in the Savannah River locality. Water and HTO were extracted by freeze-drying, and organically bound T was estimated following combustion of the dried needles. The HTO content of the atmospheric water vapour was also measured. Figure 4.5 shows Sweet & Murphy's results converted from pCi per ml of water to TU. All three variables declined approximately in proportion to distance from the plant, as would be expected for long-term releases. By extrapolation, T in atmospheric water vapour would fall to background (about 13 TU – Fig. 4.2) at a distance of 100 km.

The activity had accumulated over a number of years, and the organically bound tritium, having a long biological half-life, accounted for most of the tritium in the pine needle litter. Sweet & Murphy found evidence of direct conversions of HT to organically bound tritium in pine needles. The HTO in the needles, and probably most of the HTO in the air in the locality, derived from oxidation of HT in the soil.

Fig. 4.5. Tritium in environment of Savannah River Plant.

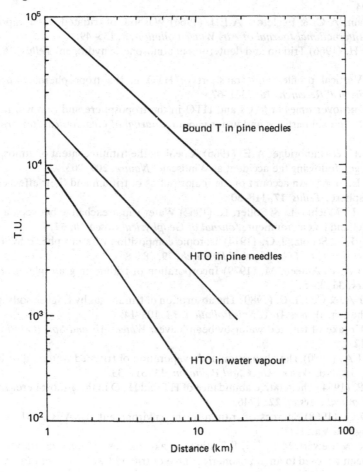

References

Amano, H. & Kasai, A. (1988) The transfer of atmospheric HTO released from nuclear facilities during normal operation. *Journal of Environmental Radioactivity*, **8**, 239–53.

Begemann, F. & Libby, W.F. (1957) Continental water balance, ground water inventory and storage times, surface ocean mixing rates and worldwide water circulation patterns from cosmic ray and bomb tritium. *Geochemica Cosmochemica Acta*, **12**, 277–96.

Belot, Y., Gauthier, D., Camus, H. & Caput, C. (1979) Prediction of the flux of tritiated water from air to plant leaves. *Health Physics*, **37**, 575–83.

Booker, D.V. (1965) Exchange between water droplets and tritiated water vapour. *Quarterly Journal Royal Meteorological Society*, **91**, 73–9.

Brutsaert, W.H. (1982) *Evaporation into the Atmosphere*. Dordrecht: D. Reidel.

Butler, H.L. & Le Roy, J.H. (1965) Observation of biological half-life of tritium. *Health Physics*, **11**, 283–5.

Chamberlain, A.C. (1981) Emission of fission products and other activities during the accident to Windscale pile No. 1 in October 1957. AERE Harwell, Oxon, Report M3194.

Chamberlain, A.C. & Eggleton, A.E.J. (1964) Washout of tritiated water vapour by rain. *International Journal of Air, Water Pollution*, **8**, 135–49.

Ehhalt, D.H. (1966) Tritium and deuterium in atmospheric hydrogen. *Tellus*, **18**, 249–55.

(1971) Vertical profiles and transport of HTO in the troposphere. *Journal of Geophysical Research*, **76**, 7351–67.

(1973) Turnover times of ^{137}Cs and HTO in the troposphere and removal rates of natural aerosol particles and water vapour. *Journal of Geophysical Research*, **78**, 7076–86.

Ehhalt, D.H. & Bainbridge, A.E. (1966) A peak in the tritium content of atmospheric hydrogen following the accident at Windscale. *Nature*, **209**, 903.

Eriksson, E. (1965) An account of the major pulses of tritium and their effects in the atmosphere. *Tellus*, **17**, 118–30.

Friedman, I., Machta, L. & Soller, R. (1962) Water vapor exchange between a water droplet and its environment. *Journal of Geophysical Research*, **67**, 7.

Friedman, I. & Scholz, T.G. (1974) Isotopic composition of atmospheric hydrogen, 1967–9. *Journal of Geophysical Research*, **79**, 785–8.

Garland, J.A. & Ameen, M. (1979) Incorporation of tritium in grain plants. *Health Physics*, **36**, 35–8.

Garland, J.A. & Cox, L.C. (1980) The absorption of tritium gas by English soils, plants and the sea. *Water, Air & Soil Pollution*, **14**, 103–14.

(1982) Uptake of tritiated water by bean leaves. *Water, Air and Soil Pollution*, **17**, 207–12.

Garland, J.A. (1980) The absorption and evaporation of tritiated water vapor by soil and grassland. *Water, Air & Soil Pollution*, **13**, 317–33.

Harteck, P. (1954) The relative abundance of HT and HTO in the atmosphere. *Journal of Chemical Physics*, **22**, 1746.

Jacobs, D.G. (1968) Sources of tritium in the environment. USAEC, TID – 24635 Springfield Va., NTIS.

Kline, J.R. & Stewart, M.L. (1974) Tritium uptake and loss in grass vegetation which has been exposed to an atmospheric source of tritiated water. *Health Physics*, **26**, 567–73.

Mason, A.S. & Ostlund, H.G. (1979) Atmospheric HT and HTO. Distribution and large scale circulation. In *Behaviour of Tritium in the Environment*, pp. 3–15. Vienna: IAEA.

Moghissi, A.A., Bretthauer, E.W. & Patzer, R.G. (1987) Biological concentration of ^{3}H. *Health Physics*, **53**, 385–8.

Murphy, C.E. & Prendergast, M.M. (1979) Environmental transport and cycling of tritium in the vicinity of atmospheric releases. In: *Behaviour of Tritium in the Environment*, pp. 361–72. Vienna: IAEA.

Murphy, C.E., Watts, J.R. & Corey, C.E. (1977) Environmental tritium transport from atmospheric release of molecular tritium. *Health Physics*, **33**, 325–51.

National Council on Radiation Protection and Measurements (NCRP) (1979) Tritium in the environment. Washington, DC.

Osborne, R.V. (1966) Absorption of tritiated water vapour by people. *Health Physics*, **12**, 1527–37.

(1972) Permissible levels of tritium in man and the environment. *Radiation Research*, **50**, 197–211.

Ostlund, H.G. & Fine, R.A. (1979) Oceanic distribution and transport of tritium. In *Behaviour of Tritium in the Environment*, pp. 303–12. Vienna: IAEA.

Rutter, A.J. (1975) The hydrological cycle in vegetation. In *Vegetation and the Atmosphere*, vol. 1, ed. J.L. Monteith, pp. 133–54. New York: Academic Press.

Schell, W.R., Sanzay, G. & Payne, B.R. (1974) World distribution of environmental tritium. In *Physical Behaviour of Radioactive Contaminants in the Atmosphere*, pp. 396–400. Vienna: IAEA.

Schmidt, U. (1974) Molecular hydrogen in the atmosphere. *Tellus*, **26**, 78–90 and (*Errata*) **27**, 93–4.

Sepall, O. & Mason, S.G. (1960) Vapour/liquid partition of tritium in tritiated water. *Canadian Journal of Chemistry*, **23**, 2024–5.

Slinn, W.G.N., Hasse, L., Hicks, B.B., Hogan, A.W., Lal, D., Liss, P.S., Munnich, K.O., Sehmel, G.A. & Vittori, O. (1978) Some aspects of the transfer of atmospheric trace constituents past the air–sea interface. *Atmospheric Environment*, **12**, 2055–87.

Sweet, C.W. & Murphy, C.E. (1981) Oxidation of molecular tritium by intact soils. *Environmental Science & Technology*, **15**, 1485–7.

(1984) Tritium deposition in pine trees and soil from atmospheric releases of molecular tritium. *Environmental Science & Technology*, **18**, 358–61.

United Nations Scientific Committee on Effects of Atomic Radiations (1977) *Ionising Radiation: Sources and Biological Effects*. New York: United Nations.

Weiss, W., Roether, W. & Dreisigacker, E. (1979) Tritium in the North Atlantic Ocean; Inventory input and transfer into deep water. In: *Behaviour of Tritium in the Environment*, pp. 315–35. Vienna: IAEA.

5

○ ○

Plutonium

5.1   Formation of plutonium aerosols

Plutonium is a man-made element, and only infinitesimal traces occur naturally. It melts at 641°C and boils at 3330°C. ^{239}Pu is formed in nuclear reactors by neutron capture in ^{238}U, followed by two successive beta decays (Fig. 5.1). Further neutron captures lead to ^{240}Pu and ^{241}Pu. ^{238}Pu is formed from ^{239}Pu by (n,2n) reactions, or from ^{235}U by three successive neutron captures and two beta decays. Table 5.1 shows the half-lives, alpha and X-ray energies of the principal Pu isotopes.

Plutonium aerosols can be formed in various ways, including:

(a)   Oxidation or volatilisation of Pu metal.

(b)   Oxidation or volatilisation of irradiated U or UO_2.

(c)   Droplet dispersion from aqueous solutions or suspensions of Pu.

(d)   Resuspension of soil or dust which has become contaminated with Pu.

The particle size of Pu aerosols is very variable, depending on the mode of formation. In Fig. 5.2, curves A, B and C show size spectra obtained by Carter & Stewart (1971) in laboratory experiments on the oxidation of Pu metal in air. In controlled oxidation at temperatures below the ignition point (about 500°C), scaly, friable, oxide particles were produced, with median diameter increasing with temperature. Few particles less than 1 μm in diameter were found. When the delta alloy of Pu was used, the oxide was more adherent, and the particle size larger. Increase of particle size with increase of temperature was also found in laboratory oxidation of uranium metal (Megaw *et al.*, 1961), and was ascribed to sintering of the oxide layer.

Quite different results were obtained when Pu metal was heated in argon above its melting point, and droplets of molten metal were allowed to fall down a column in air. The vigorous oxidation raised the temperature of the drops sufficiently to generate Pu vapour which condensed as a fume, comprising particles of about 0.1 μm diameter aggregated in chains. Fume was also generated when an electrical current was passed through a plutonium wire (Brightwell & Carter, 1977), giving the particle size distribution shown in curve D of Fig. 5.2.

Fig. 5.1. Formation of Pu isotopes.

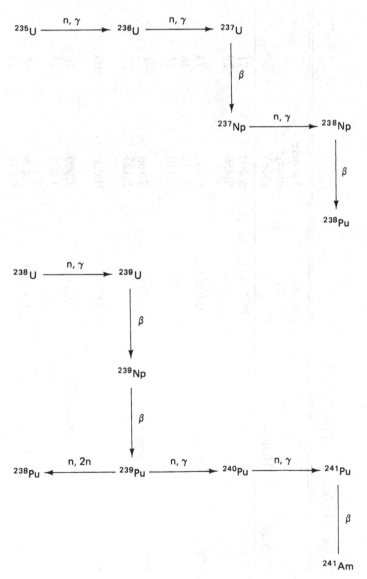

Table 5.1. *Isotopes of plutonium*

Nuclide	Half-life	Specific activity (Ci g^{-1})	Particle energy (MeV)	%	X-ray (MeV)	%
^{238}Pu	86.4 a	17.55 (α)	5.498 5.456	71 29	0.017	10.8
^{239}Pu	24400 a	0.062 (α)	5.155 5.143 5.110	73 16 11	0.017	4.6
^{240}Pu	6600 a	0.23 (α)	5.168 5.123	75 25	0.017	10.3
^{241}Pu	13.2 a	113 (β)	0.020 (β)	100		
^{242}Pu	3.8×10^5 a	0.004 (α)	4.900 4.856	76 24	0.017	8.6
^{241}Am	458 a	3.45 (α)	5.486 5.443	85 13	0.018	37

Plutonium is so toxic that processing and fabrication are always done in sealed cells or glove boxes, but accidental dispersions of aerosol occur from time to time. Following combustion of Pu metal chips in a production area at Rocky Flats, Colorado, in 1964, airborne contamination was widespread. Alpha tracks from individual particles caught on membrane filters were detected on nuclear film, and the Pu content, and hence the particle size, was deduced (Fig. 5.2, curve E). The activity median diameter was 0.3 μm (Mann & Kirchner, 1967). The same method, used during normal operations in a production area at Los Alamos, gave activity median diameters in the range 0.15 to 0.65 μm (Moss *et al.*, 1961). However, when a spill occurred, followed by clean-up operations, the Pu particles were found to be associated with inert dust particles of mass median diameter 7 μm.

Fig. 5.2. Particle size distributions of Pu oxidising in air at (A) 20°C, (B) 123°C, (C) 450°C (Carter & Stewart, 1971). D, Pu fume from exploded wire (Brightwell & Carter, 1977). E, Pu fume from fire (Mann & Kirchner, 1967).

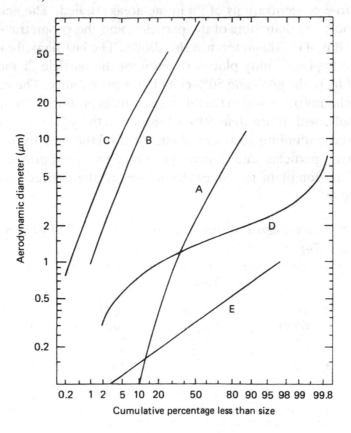

Sherwood & Stevens (1965) examined glass-fibre filters from personal air samplers worn by workers in the Radiochemical Laboratories at Harwell. The filters were mounted in an Araldite mixture which rendered them transparent and were covered by autoradiographic stripping film. After exposure and development, the samples were viewed with a high-power optical microscope. Particles were sized, and their activity determined from the number of alpha tracks coming from them. An extremely wide range of particle sizes, 0.2 to 90 μm, was found. The smaller particles were plutonium compounds or alloys, and the larger were inert particles with one or more small Pu particles attached to them. An example of the latter is shown in Fig. 5.3.

Elder *et al.* (1974) obtained a statistical distribution of aerosol particle sizes by operating multi-stage cascade impactors (Andersen samplers) in the ventilation ducts at three plutonium processing plants. When the size distribution was plotted on log-probability paper, a reasonable approximation to a straight line was obtained with about 80% of the samples, but some showed bimodal distributions. Table 5.2 shows the mean airborne concentrations of Pu in the areas studied. The activity aerodynamic mean diameters of the particles, and the geometric standard deviations of the diameters are also shown. The last-named can be read off the log-probability plot as the ratio of the particle diameters corresponding to the 84% and 50% cumulative probability. The grinding and machining processes in the fabrication areas produced relatively large aerosols, with more than 50% of the Pu activity in the 1–5 μm range. In the plutonium recovery plant, 70% of the activity was in submicrometre particles, and this area presented the worst problems in designing filtration plant to reduce the effluent to the desired level of 2×10^{-3} Bq m^{-3}.

Table 5.2. *Plutonium aerosols in ventilation ducts of USAEC plants (Elder et al., 1974)*

Location	Major operations	Mean airborne activity (Bq m^{-3})	AAMD (μm)	σ_g
A	R & D	2×10^3	1 to 3	1.5 to 3
B	R & D	2×10^4	1 to 4	1.5 to 3.5
C	Fabrication	1×10^4	3 to 5	1.5 to 2.5
D	Recovery	1×10^5	0.1 to 1	1.5 to 4
E	Fabrication	2×10^4	2 to 4	1.5 to 3

Sanders & Boni (1980) collected plutonium-bearing particles from exhaust systems at the Savannah River Processing Plant, and counted alpha tracks generated by them in nuclear emulsions. They also irradiated the particles with thermal neutrons (9×10^4 cm^{-2}) and counted the number of fission tracks produced. By comparing the numbers of alpha and fission fragment tracks, it was possible to distinguish low-irradiated from high-irradiated plutonium, which has an increased proportion of the isotope ^{238}Pu (Fig. 5.1) giving more alpha tracks per Bq of ^{239}Pu.

Sanders & Boni used a photo-microscope with reflected light to measure the sizes of Pu-bearing particles, and these were found to

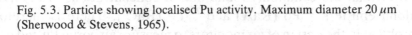

Fig. 5.3. Particle showing localised Pu activity. Maximum diameter 20 μm (Sherwood & Stevens, 1965).

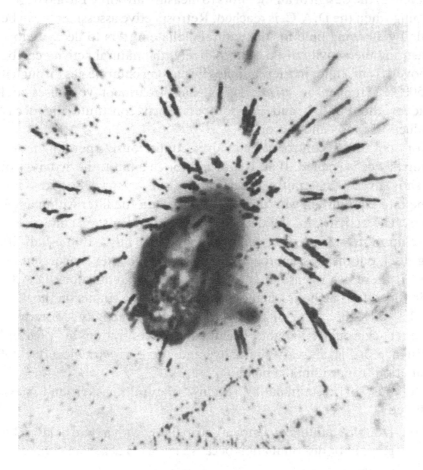

follow log-normal distributions, with geometric mean optical diameters varying from 4 to 12 μm at various locations. The particles were also subjected to microprobe analysis to determine the chemical composition, which was found to be similar to crustal material. Submicrometric fume particles of Pu were not reported.

5.2   Detection of airborne plutonium

The Derived Air Concentration (D.A.C.) is the maximum permissible concentration for occupational workers. It is equal to the Annual Limit of Intake (A.L.I.) divided by the volume of air assumed to be inhaled during working hours per year, namely 2.4×10^3 m^3, (International Commission on Radiological Protection, 1979). The D.A.C. of ^{239}Pu is 0.09 Bq m^{-3}, which is about two orders of magnitude lower than the commonly found indoor concentrations of the natural alpha emitters ^{218}Po (RaA) and ^{214}Po (RaC'). There is therefore a problem in the design of air monitors to measure airborne Pu and to give warning when the D.A.C. is reached. Retrospective assessments can be made by allowing time for the natural alpha emitters to decay before measuring the activity of Pu. Alternatively, the natural activity can be removed from a filter in a few minutes by passing chlorine gas through it at 500°C (Stephenson *et al.*, 1971). This treatment volatilises and removes the Pb, Bi, and Po isotopes which constitute the decay products of radon and thoron.

For operational control of Pu in air, a continuously operating monitoring device is needed. If air is drawn through a filter, the activities of the short-lived alpha-emitting decay products of ^{222}Rn (^{218}Po and ^{214}Po) come into equilibrium with the rate of arrival on the filter within a few hours. The activities of the decay products of ^{220}Rn (^{212}Bi and ^{212}Po) build up more slowly, since their parent, ^{212}Pb, has a 10.6-h half-life (Fig. 1.1), but they too approach equilibrium if the filter is operated for 24 h. If there is a release of ^{239}Pu aerosol, equivalent to one D.A.C. maintained for one hour, then the additional alpha counts on the filter will be only about 1% of those due to natural activity. Moreover, natural background is variable, because the concentrations of ^{222}Rn and ^{220}Rn in outside air vary diurnally, and there is variable effect of emanation from within the building.

Methods of discriminating against natural background have included:

(a)   Double counting of a moving strip filter, with a time delay. This gives poor discrimination because natural background varies.

(b) Alpha-beta anti-coincidence, relying on the nearly simultaneous alpha and beta emissions from the parent/daughter
pairs ^{214}Bi/^{214}Po and ^{212}Bi/^{212}Po. This does not provide discrimination against ^{218}Po.

(c) Size selection, using an impaction device. This relies on Pu
aerosol particles being larger, and particles carrying natural
activity smaller, than a certain aerodynamic diameter (e.g. 0.3
μm).

(d) Alpha spectrometry, relying on the fact that the alpha energy of
^{239}Pu, and other actinide isotopes except ^{242}Cm, is less than the
alpha energy of the ^{222}Rn and ^{220}Rn decay products.

Alpha spectrometry is superseding the other methods. Absorption of
alpha energy in the material of the filter, and in the air gap (which is
normally about 5 mm) between the filter and the detector, broadens the
apparent alpha energy spectrum from each nuclide. Figure 5.4 shows a
typical alpha energy spectrum from natural activity collected in a filter
through which laboratory air had been drawn at 30 l min^{-1} for 24 h
(Ryden, 1981). Also shown is the peak produced by one D.A.C. of
^{239}Pu present in air for 1 h (0.08 Bqh m^{-3}). The low-energy tail from the
combined peaks of the 6.0 MeV alpha from ^{218}Po and the 6.05 MeV
alpha from ^{212}Bi extend into the channels which register the 5.15 MeV
alpha from ^{239}Pu. Consequently, the latter contributes only a small
bulge on the combined spectrum. To overcome this, Ryden (1981)
developed an instrument (Harwell T 3280) incorporating an information processor. The counts in the groups of channels marked C_1, C_2,
C_3 in Fig. 5.4 are registered and used to calculate the contribution of
natural activity to the signal group of channels marked S, centred on the
^{239}Pu peak. In this way, the ^{239}Pu counts are discriminated.

There remains the problem of statistical accuracy, since the total
count in the ^{239}Pu peak, assuming a dosage of 0.008 Bqh m^{-3} and a
counter efficiency of 32%, is less than 3 cpm. A statistical problem of
another kind lies in the fact that the D.A.C. of ^{239}Pu in air is equivalent to one PuO$_2$ particle of diameter 1.6 μm per m^3 of air. At a
sampling rate of 30 l min^{-1}, one such particle would be sampled in half
an hour.

To separate Pu from biological materials, and from soil, acid leaching
is followed by ion exchange and electroplating onto platinum discs for
alpha spectrometry (Chu, 1971). The alpha energies of ^{239}Pu and ^{240}Pu
are so close that the isotopes can only be distinguished by mass

Fig. 5.4. Alpha spectrum of ^{239}Pu and natural activities.

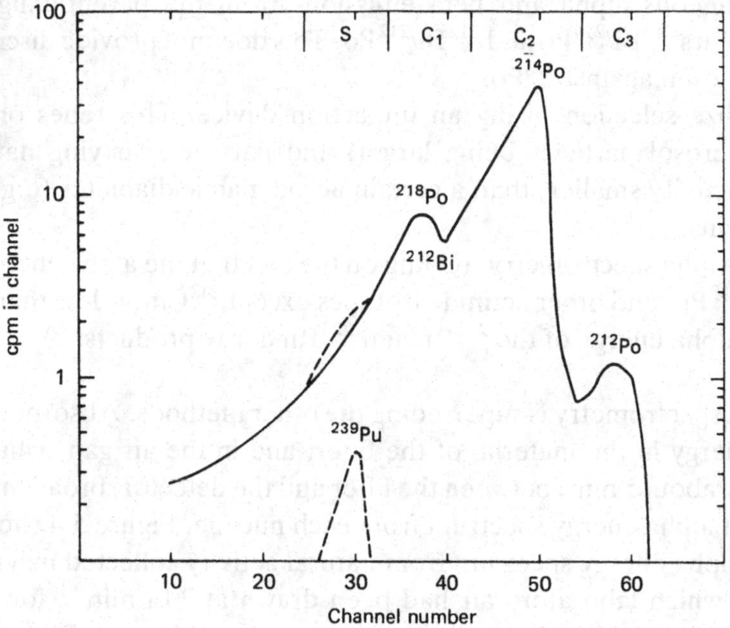

spectrometry. Consequently, the combined activity is often reported as $^{239+240}$Pu.

5.3   Plutonium in the environment

Plutonium has been dispersed globally by weapon tests and by the burn-up of the S.N.A.P. satellite in the upper atmosphere. There have also been local operational and accidental dispersions (Harley, 1980). Some of the major events are discussed below.

(i) Thermonuclear explosions
In thermonuclear weapons, neutrons are absorbed in a blanket of ^{238}U, where they induce fission and thus increase the power of the explosion. Neutron capture in ^{238}U also produces ^{239}Pu, ^{240}Pu and ^{241}Pu. About 15 PBq (400 kCi) of Pu isotopes were disseminated by atmospheric testing of thermonuclear weapons, with a peak period in 1961–2 (Hardy *et al.*, 1973). Most of the activity was carried into the stratosphere by the heat of the explosion. At ground level, there was a seasonal variation in the air concentration, similar to the seasonal variation in bomb-derived fission products and tritium (Fig. 4.1), with peaks in early summer,

when downwards transport from stratosphere to troposphere is most effective. This is well illustrated by the results of Perkins & Thomas (1980) shown in Fig. 5.5. After the cessation of atmospheric tests by the major powers in 1963, the amount of Pu in the stratospheric reservoir declined with a residence half-life of about 10 months. Figure 5.6 shows annual average $^{239+240}$Pu in air at ground level in the USA (Harley, 1980) and the UK (Cambray *et al.*, 1985). The concentrations have fallen by a factor 1000 from the 1963 peak. The recent fluctuations are the result of atmospheric tests in China.

The first US thermonuclear bomb was detonated near the ground at Bikini atoll in 1954, and much surface material was incorporated into the fireball, but most US and USSR thermonuclear tests were conducted at altitude, and relatively small amounts of material were vapourised. Consequently small particles were formed on condensation, and these have become attached to the general stratospheric aerosol (Harley, 1980).

Fig. 5.5. ^{238}Pu and ^{239}Pu in surface air at Richland, Washington (Perkins & Thomas, 1980).

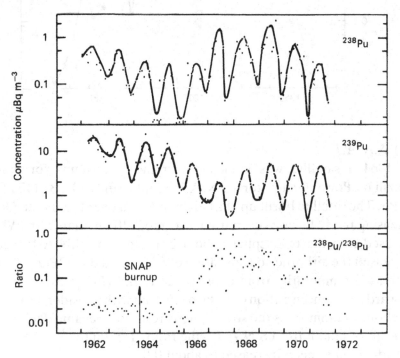

Fig. 5.6. Concentration of $^{239+240}$Pu in air: A, UK; B, USA – New York; C, Rocky Flats, site 1; D, Rocky flats, site 4; E, Eskmeals.

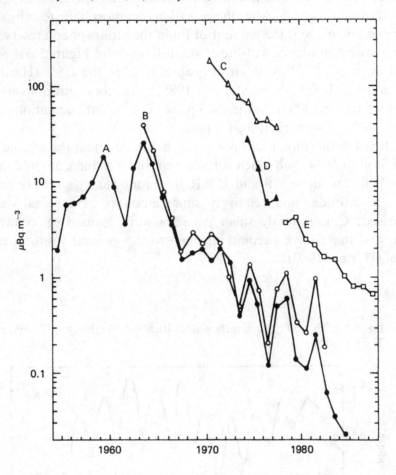

(ii) S.N.A.P.

In 1964 a satellite was launched carrying a 'Systems for Nuclear Auxiliary Power Generator' (S.N.A.P.) containing 630 TBq (17 kCi) of ^{238}Pu. The satellite burnt up at a height of 50 km over the Indian Ocean, releasing the ^{238}Pu as oxide particles of about 10 nm diameter. When it diffused into the troposphere about 2 a later, the S.N.A.P. activity increased the airborne concentration of ^{238}Pu, and the ^{238}Pu/^{239}Pu ratio increased temporarily from about 0.024 to 0.5 (Fig. 5.5). From 1967 onwards the concentration of ^{238}Pu decreased with a residence half-time of about 14 months, as the stratospheric reservoir became depleted. As a result of fallout from S.N.A.P., the ratio of ^{238}Pu to $^{239+240}$Pu in soil world-wide has been increased to about 0.05.

(*iii*) *Low-yield weapon tests*

In some tests carried out at the Nevada Test Site in 1956–8, fission was incomplete and some plutonium was dispersed. Also some weapons were exploded chemically without nuclear fission in safety tests. Analysis of soil from an area extending about 500 km north-eastwards from the N.T.S. into the neighbouring state of Utah showed excess Pu over the amounts expected from global fallout (Hardy, 1976). The Pu from these low-level, low-yield tests was distinguished from global fallout by two criteria:

(a)   On account of the incomplete fission, the activity ratio $^{239+240}$Pu/^{137}Cs was higher than the ratio (0.016) typical of global fallout.

(b)   Mass spectrometry showed that the ^{240}Pu/^{239}Pu atom ratio in the Nevada source was 0.05, which is typical of weapons grade Pu. The neutron fluxes in thermonuclear detonations cause neutron capture in ^{239}Pu, and the atom ratio of ^{240}Pu/^{239}Pu in global fallout is 0.18.

The total fallout from the low-yield trials in Nevada and Utah (outside the NTS boundary) was estimated by Hardy at 26 TBq (700 Ci), compared with about 9 TBq of global $^{239+240}$Pu falling in the same area. The peak measured fallout, outside the N.T.S., was 600 Bq m^{-2} at Eureka, NV. This compares with mean global fallout of 60 Bq m^{-2} in the USA (Hardy, 1976) and 50 Bq m^{-2} (normalised to 1000 mm a^{-1} rainfall) in the UK (Peirson *et al.*, 1982). Within the N.T.S., the deposition of Pu was much greater, ranging from 4×10^4 to 2×10^8 Bq m^{-2} (Martin & Bloom, 1980).

The particle size of the Pu disseminated in the N.T.S. trials is not known, but it was probably relatively large. Hardy noted that replicate samples of soil analysed for Pu gave unusually disparate results, probably on account of sampling variability related to particle size.

The conditions in Nevada are favourable for resuspension of Pu from the ground. Because the area round the N.T.S. is arid, Pu has not been moved down the soil profile by leaching or by cultivation, and more than 50% of the Pu was found to be in the top 20 mm of soil about 15 a after deposition (Anspaugh *et al.*, 1975). The mechanisms of resuspension of particles from the ground are considered in a later chapter. The resuspension factor K_r is defined:

$$K_r = \frac{\text{airborne concentration (units/m}^3\text{)}}{\text{surface concentration (units/m}^2\text{)}}$$

Measurements at the N.T.S. by Anspaugh *et al.* (1975) gave a mean resuspension factor of 10^{-9} m^{-1}. When the soil is dry, and vegetation is scarce, K is likely to be greater than normal. If the density of a sandy soil is taken as 2000 kg m^{-3}, 50% of the Pu is in the top 20 mm, and there is a dust loading of 80 μg m^{-3} in the air from blowing soil, then a value of $K_r = (80 \times 10^{-6})/(8 \times 10^4) = 10^{-9}$ m^{-1} can be expected.

Shinn *et al.* (1989) have investigated the removal of plutonium from surface soils at the N.T.S. by use of a vacuum truck of the type used to clean streets. Vegetation on the site comprised small shrubs giving 15% ground cover. This was cleared and two to five cm of soil was then removed by the vacuum truck.

This treatment removed 92% of the Pu contamination, as measured by the activity of the ^{241}Am daughter product of ^{241}Pu. It was even more effective in reducing the subsequent rate of re-suspension by the wind, which was 92% lower than before the treatment. As discussed later (Section 6.13), re-suspension depends strongly on the shearing stress exerted by the wind on the ground, and this was less after the treatment because the aerodynamic roughness of the surface was reduced.

(iv) Rocky Flats

Beginning in 1964, cutting oil containing Pu leaked from a storage area at the Rocky Flats processing plant, near Denver, Colorado, and contaminated the surrounding soil. Resuspension by wind spread the contamination. The low ^{240}Pu/^{239}Pu atom ratio distinguished the local source from global fallout, and enabled the isopleths of deposition to be established (Krey, 1976).

Krey plotted the area within the isopleth (Fig. 5.7), and found that the area (km^2) within the isopleths, A_p (Bq m^{-2}), was proportional to $A_p^{-0.824}$. Since the area increased less rapidly than A_p^{-1}, a summation was possible, and this gave a result 0.15 TBq (4 Ci) as the total deposit of Pu from the spill at Rocky Flats. The highest deposit was 7.4×10^4 Bq m^{-2} (2 μCi m^{-2}). This greatly exceeded the peak off-site contamination at N.T.S., but the area involved was much less, as would be expected from the modes of dispersal in the two cases.

As Rocky Flats is only 25 km from Denver, and lies in an area of increasing population, concern was felt about airborne Pu derived from resuspension of contaminated soil. Monitoring for airborne Pu was done at sites 1 and 4, lying 200 and 800 m respectively from the spill in the direction of the prevailing wind (Volchok *et al.*, 1977). The results are compared in Fig. 5.6 with the airborne concentration at New York, due to global fallout.

Ten years after the spread of activity, about two-thirds of the deposited Pu was in the top 50 mm of soil and reduced availability for resuspension probably accounts for the fall in air concentrations shown in Fig. 5.6. The region is moderately arid, with an annual rainfall of 400 mm, and the area near the source of Pu is untilled grassland. Because the Pu in air was mostly attached to soil particles, the particle size was large, and only about 25% was in the respirable fraction (Volchok *et al.*, 1972).

(v) Palomares

Two plutonium weapons fell from a US aircraft after a collision over Palomares, Spain, in 1966 and chemical explosions occurred on impact with the ground. The plutonium was converted into a fine oxide aerosol and spread by wind (Harley, 1980). Soil samples were analysed, and

Fig. 5.7. Areas within isopleths of Pu in soil: ○, Rocky Flats; △, Palomares; □, Sellafield.

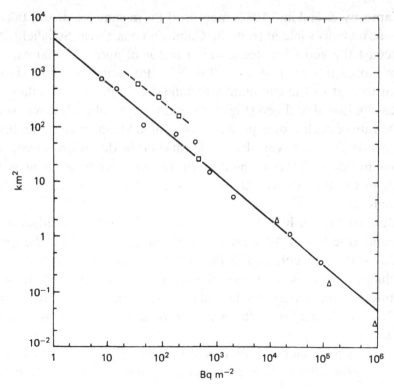

showed that an area of 2 ha was contaminated with more than 1.2×10^6, 16 ha with more than 1.2×10^5, and 200 ha with more than 1.2×10^4 Bq m^{-2}. The data, plotted in Fig. 5.7, lie close to the line of Krey's (1976) results from Rocky Flats. This may be fortuitous, but it may be significant that redispersal of soil by wind was a factor in the spread of contamination at Palomares as at Rocky Flats. Pu was removed from the most severely contaminated area at Palomares by removing the topsoil, and elsewhere it was buried by ploughing.

(vi) Sellafield (formerly Windscale)
Between 1957 and 1983 about 700 TBq of plutonium were discharged to sea by pipeline from the Sellafield reprocessing plant in Cumbria. Since 1980, the annual rate of discharge has been reduced. The ^{238}Pu/$^{239+240}$Pu ratio has increased from about 0.05 in the late 1950s to 0.3 in recent years, as more highly irradiated fuel has been processed. The ratio in the cumulative emissions is about 0.15, and this provides one criterion for distinguishing Sellafield Pu in the environment from global fallout, in which the ^{238}Pu/$^{239+240}$Pu ratio is about 0.04 (Hardy *et al.*, 1973).

Cambray & Eakins (1982) measured Pu in samples of soil taken at various distances inland from the Cumbrian coast near Sellafield. Near the coast the activity exceeded by a factor of about 10 that expected from global fallout, and the ^{238}Pu/$^{239+240}$Pu ratio was approximately equal to that of the cumulative Sellafield effluent. Going inland, the excess Pu in soil declined (Fig. 5.8), and the rate of decline was similar to the rate of decline of deposits of NaCl inland from coasts (Cambray & Eakins, 1982). However, the Pu/Na ratio in the deposit was two orders of magnitude higher than the ratio in seawater, whereas the activity of ^{137}Cs in the deposit was about as expected from the Cs/Na ratio in seawater.

Most of the Pu in seawater is not in solution but is adsorbed on suspended sediment. When waves break at the sea surface, the spray is enriched in sediment, and in Pu, relative to the bulk seawater, by a mechanism which is the basis of the technique of froth flotation used by chemical engineers to extract small particles from aqueous suspensions. ^{137}Cs, being in solution in the seawater, is not enriched relative to Na in the spray.

From measurements of Pu in soil in the coastal regions of Cumbria (Peirson *et al.*, 1982), the areas contaminated to various levels, in excess of global fallout, can be calculated, and the results are plotted in Fig. 5.7. At low levels of deposit, to the left of the graph, the slope is similar to that characterising the Rocky Flats and Palomares deposits. There is a cut-off in the Cumbria data at about 500 Bq m^{-2}, and this is related to the fact that the source is disperse (i.e. in the coastal waters) whereas at Rocky Flats and Palomares it was localised to a small area. Graphical integration shows that about 0.06 TBq (1.7 Ci) of Pu, in excess of global fallout, has been deposited on land in Cumbria within the 37 Bq m^{-2} (1 mCi km^{-2}) isopleth. Assuming that the logarithmic slope 0.824 in Fig. 5.7 applies to lower levels of deposition, there is a further 0.13 TBq beyond the 37 Bq m^{-2} isopleth, giving a total areal deposition of 0.2 TBq (5.4 Ci). Thus the total areal/deposition ratios of Pu near Sellafield and near Rocky Flats are of similar magnitude.

Since 1977, the airborne activity of Pu has been measured at a site at Eskmeals, near the Ravenglass estuary and 11 km SSE of Sellafield (Cambray *et al.*, 1987). Figure 5.6 shows that the concentrations, though declining by a factor of five from 1979 and 1985, were about one order of magnitude higher than the contemporary levels in the UK or USA generally. However, the Eskmeals concentrations were much lower than the world-wide peak concentrations from weapons testing in the 1960s.

Fig. 5.8. Pu in soil inland from coast of Cumbria: O, $^{239+240}$Pu; $\triangle$, ^{238}Pu.

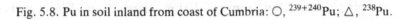

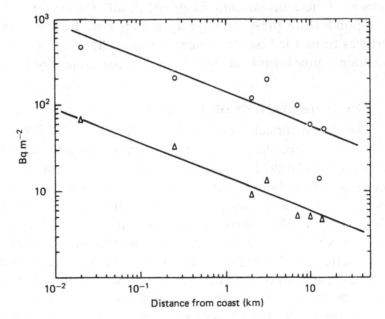

Fig. 5.9. $^{239+240}$Pu in air at Eskmeals ($\triangle$) and in silt at Newbiggin ($\bigcirc$).

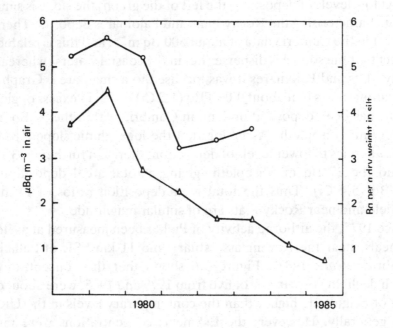

Hunt (1985) has measured the concentration of Pu in coastal silt at Newbiggin, about 1 km from the Eskmeals site, and Fig. 5.9 shows a comparison of the concentrations in air and in silt. The observed Pu in air corresponds to the presence of $0.6\,\mu g$ of silt per m^3. Resuspension of silt particles from dried-out areas near the shore line may contribute, but the main source is probably spray from the surf zone offshore.

5.4   The inhalation of plutonium

The International Commission on Radiological Protection (1987) has reported on the deposition of plutonium particles in the lung and the transportability to other tissues. There is a lack of human data linking uptake to airborne concentration of plutonium, so extrapolations have to be made from animal data and from human inhalation experiments with other particles. Experiments with lead aerosols, described below (Fig. 7.10) suggest that 20–80% of Pu inhaled as submicrometre particles might be deposited in the lung, depending on particle size and duration of the breathing cycle. For a plutonium aerosol with an activity median aerodynamic diameter of 1 μm, breathed through the nose, ICRP (1987) assumes 25% deposition in the pulmonary region and 63% deposition in the whole respiratory tract.

Plutonium oxide is very insoluble, and particles are removed slowly from the lung, mainly by phagocytosis. The crystalline structure of the oxide particles, which depends on the temperature of formation, appears to affect transportability, PuO_2 formed at higher temperatures being least transportable. From animal experiments, it appears that PuO_2 particles in the nanometric size range are more transportable than those in the micrometre size range (ICRP, 1987).

The recommendations of ICRP (1987), which are based on the ICRP lung model, adopt a 500-d clearance half-time of insoluble Pu compounds from the pulmonary region of the lung. Of the material cleared, 15% goes to the pulmonary lymph nodes. In the lymph nodes, a residence half-time of 1000 days is assumed to apply to 90% of the activity, the remaining 10% staying indefinitely. Newton *et al.* (1983) measured the intensity of 13–20 keV X-rays from ^{238}Pu in the lungs of a subject who had inhaled the activity accidentally, and found that it was cleared with an effective half-life of 750 d, almost entirely by systemic uptake.

Bennett (1976) used the measured airborne activity of $^{239+240}$Pu in New York (Fig. 5.6) to calculate the resulting activity in various organs of the body. For the years 1954–63, when no measurements of airborne Pu were available, he estimated values from the fallout rates which were measured. He assumed a breathing rate of 20 m^3 d^{-1}, 32% deposition of Pu aerosol in the pulmonary region of the lung, and a retention half-life of 500 d applying to 60% of the deposited activity, the other 40% being eliminated quickly. Of the Pu deposited, 2.75% was assumed to be transferred to bone, where it was retained with a half-life of 40 a. Figure 5.10 shows the results of Bennett's calculations of the burden of Pu in the lung and bone, and also the burden in lung assuming the retention half-life is 2 a instead of 500 d. As the air concentration has fallen in the last 20 a, nearly all the Pu in the skeleton of mature adults derives from activity inhaled in the 1960s. In the peak year, 1963, the average intake of $^{239+240}$Pu by inhalation was 0.4 Bq. This compares with an annual limit of intake by the general public of 9 Bq (Haywood, 1987).

Also shown in Fig. 5.10 are the means and one s.d. ranges (based on log-normal distributions) of Pu in autopsy samples of UK subjects dying in 1980–4 (Popplewell *et al.*, 1985). The agreement with the calculated levels is good, especially if the retention half-life in the lung is taken as

700 d. Little difference would be expected in the Pu burdens in the UK and USA since the air concentrations were similar.

Popplewell *et al.* (1985) also measured Pu in autopsy samples from persons who had lived in West Cumbria, and found levels averaging about twice those from other regions of the UK. Mass spectrometry showed lower activity ratios $^{240}Pu/^{239}Pu$ and $^{241}Pu/^{239}Pu$ in these subjects than in the general population (Popplewell *et al.*, 1989). In four of the subjects, the $^{240}Pu/^{239}Pu$ ratio was measured in both lung and liver, and found to be lower in the lung. It was concluded that the enhanced concentrations derived from the plutonium processed in the early days of the Sellafield plant had a low $^{240}Pu/^{239}Pu$ ratio, and it was concluded that inhalation of plutonium from this source caused the enhanced concentrations.

5.5   Entry of plutonium into diet

Plutonium is poorly absorbed by plants from the soil, and is not translocated between plant tissues. Romney *et al.* (1981) found concentration ratios (Pu in vegetation divided by Pu in soil, both per unit dry weight) of less than 10^{-4} for leaves and less than 10^{-6} for cereal grains.

Fig. 5.10. Uptake of Pu by general population: A, calculated burden of Pu in skeleton; B_1, B_2, calculated burden in lung, 700 d and 500 d half-life. Boxes show mean and s.D. of measurements on autopsy samples.

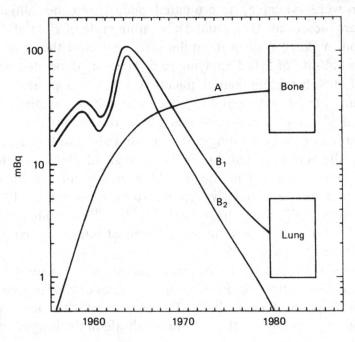

Plutonium is deposited on plant tissues by fallout and by resuspension from soil. Pinder et al. (1985) and Pinder & McLeod (1988) measured ^{238}Pu and $^{239+240}$Pu in topsoil and in corn and sunflower foliage near the Savannah River Plant. Because the ratio of Pu isotopes in contemporary fallout was different from that in soil, Pinder et al. were able to estimate how much of the Pu on the foliage derived from the soil. They found that, at harvest, resuspended soil on foliage amounted to about 0.8 g per m^2 of ground area.

After correcting for the contribution from resuspended soil, and normalising the results to unit rate of fallout, Pinder et al. found Normalised Specific Activities for ^{238}Pu on wheat, soybean, corn and tobacco foliage in the range 18–41 m^{2}d kg^{-1}. These results are similar to those found when ^{90}Sr or ^{137}Cs is deposited on grass in good growing conditions (Tables 2.17, 2.18). However, measurements of ^{238}Pu in grain give NSA values less than 1 m^{2}d kg^{-1}, lower than those for ^{137}Cs (McLeod et al., 1980; Simmonds & Linsley, 1982).

Bennett (1974) measured $^{239+240}$Pu in New York diet, and estimated the annual dietary intake, 1972–4, as 0.06 Bq. The annual rate of fallout in those years was 0.6 Bq m^{-2}, so the equivalent area – the area on which fallout equalled the average dietary intake – was 0.1 m^2. Hisamatsu et al. (1987) measured Pu in Japanese diet in 1959–62 when fallout was approaching its peak and again in 1978–80 when it was much less. The results (Table 5.3) show equivalent areas averaging 0.41 m^2 in the first and 0.55 m^2 in the second period. These results are not

Table 5.3. *Fallout and dietary Pu in Tokyo*

Year	Fallout (Bq m^{-2} a^{-1})	Dietary intake (Bq a^{-1})	Ratio (m^2)
1959	3.6	0.74	0.21
1960	1.6	0.74	0.46
1961	1.4	0.52	0.37
1962	4.1	2.44	0.60
		Mean	**0.41**
1978	0.26	0.067	0.26
1979	0.15	0.044	0.29
1980	0.04	0.044	1.10
		Mean	**0.55**

significantly different, and show that dietary Pu depends on the current rate of fallout, rather than on the cumulative total. The difference in equivalent areas between Pu in New York and Japan no doubt reflects differences in dietary sources.

An equivalent area of 0.24 m^2 was deduced by comparing the dietary intake and rate of fallout of ^{210}Pb (Chamberlain, 1983). For nuclides poorly absorbed by the roots, the contamination of plants is by foliar interception (Table 2.14) and by rainsplash of activity on the soil surface (Dreicer *et al.*, 1984). The first mechanism depends on the current rate of fallout. The second may have some dependence on cumulative fallout, but on cultivated soils this will be slight.

The annual dietary intake of Pu is several times greater than the intake by inhalation, but the fractional uptake from the gut is only about 5×10^{-4} (Simmonds *et al.*, 1982), so dietary intake has not contributed greatly to the Pu burden in the general public. Simmonds *et al.* concluded that inhalation has contributed 96% of the total.

References

Anspaugh, L.R., Shinn, J.H., Phelps, P.L. & Kennedy, N.C. (1975) Re-suspension and re-distribution of plutonium in soils. *Health Physics*, **29**, 571–82.

Bennett, B.G. (1974) Fallout ^{239,240}Pu in diet. In Report HASL 306, US Dept of Energy, Springfield, Va: NTIS, pp. 115–25.

Bennett, B.G. (1976) Transuranic element pathways to man. In: *Transuranic Nuclides in the Environment*. Vienna: IAEA, pp. 367–82.

Brightwell, J. & Carter, R.F. (1977) Comparative measurements of the short term lung clearance and translocation of PuO$_2$ and mixed Na$_2$O & PuO$_2$ aerosols in mice. In: *Inhaled Particles IV*, ed. W.H. Walton, Part 1, Oxford: Pergamon, pp. 285–301.

Cambray, R.S., Pattenden, J.J. & Playford, K. (1988) Studies of environmental radioactivity in Cumbria: Part 16. Report AERE – R 12617. Harwell: UKAEA.

Cambray, R.S. & Eakins, J.D. (1982) Pu, ^{241}Am and ^{137}Cs in soil in West Cumbria and a maritime effect. *Nature*, **300**, pp. 46–8.

Cambray, R.S., Playford, K. & Lewis, G.N.J. (1985) Radioactive fallout in air and rain: results to the end of 1984. Report AERE – R 11915. Harwell, Oxon: UKAEA.

Carter, R.F. & Stewart, K. (1971) On the oxide formed by the combustion of plutonium and uranium. In: *Inhaled Particles III*, ed. W.H. Walton, vol. 2, pp. 819–38. Oxford: Pergamon.

Chamberlain, A.C. (1983) Fallout of lead and uptake by crops. *Atmospheric Environment*, **17**, 693–706.

Chu, N.Y. (1971) Plutonium separation in soil by leaching and ion exchange separation. *Analytical Chemistry*, **43**, 449–52.

Dreicer, M., Hakonson, T.E., White, G.C. & Whicker, F.W. (1984) Rainsplash as a mechanism for soil contamination of plant surfaces. *Health Physics*, **46**, 177–87.

Elder, J.C., Gonzales, M. & Ettinger, H.J. (1974) Plutonium aerosol size characteristics. *Health Physics*, **27**, 45–53.

Hardy, E. (1976) Plutonium in soil northwest of the Nevada Test Site. In: *US Department of Energy Report HASL 306*, pp. 51–76. Springfield Va: NTIS.

Hardy, E., Krey, P.W. & Volchok, H.L. (1973) Global inventory and distribution of fallout plutonium. *Nature*, **241**, 444–5.

Harley, J.H. (1980) Plutonium in the environment – a review. *Journal of Radiation Research*, **21**, 83–104.

Haywood, S.M. (1987) Revised generalized derived limits for radioisotopes of strontium, iodine, caesium, plutonium, americium and caesium. Report NRPB-GS8. National Radiological Protection Board, Chilton, Oxon.

Hisamatsu, S., Takizawa, Y. & Abe, T. (1987) Ingestion intake of fallout Pu in Japan. *Health Physics*, **52**, 193–200.

Hunt, G.J. (1985) Timescales for dilution and dispersion of transuranics in the Irish Sea near Sellafield. *The Science of the Total Environment*, **46**, 261–78.

International Commission on Radiological Protection (1979) Limits for intakes of radionuclides by workers (ICRP 30, Part 1). *Annals of the ICRP*, **2**, Nos. 3/4.

(1987) The metabolism of plutonium and related elements (ICRP 48). *Annals of the ICRP*, **16**, Nos. 2/3.

Krey, P.W. (1976) Remote plutonium contamination and total inventories from Rocky Flats. *Health Physics*, **30**, 209–14.

Mann, J.R. & Kirchner, R.A. (1967) Evaluation of lung burdens following acute inhalation exposure to highly insoluble PuO_2. *Health Physics*, **13**, 877–82.

Martin, W.E. & Bloom, S.G. (1980) Nevada Applied Ecology Group model for estimating plutonium transport and dose to man. In: *Transuranic Elements in the Environment*, ed. W.C. Hanson, pp. 459–512. US Dept of Energy, Springfield, Va: NTIS.

McLeod, K.W., Adriano, D.C., Boni, A.L., Corey, J.C., Horton, J.H., Paine, D. & Pinder, J.E. III (1980) Influence of a nuclear fuel chemical separations facility on the plutonium content of a wheat crop. *Journal of Environmental Quality*, **9**, 306–15.

Megaw, W.J., Chadwick, R.C., Wells, A.C. & Bridges, J.E. (1961) The oxidation and release of iodine-131 from uranium slugs oxidising in air and carbon dioxide. *Reactor Science & Technology*, **15**, 176–84.

Moss, W.D., Hyatt, E.C. & Schulte, H.F. (1961) Particle size studies on plutonium aerosols. *Health Physics*, **5**, 212–18.

Newton, D., Taylor, B.T. & Eakins, J.D. (1983) Differential clearance of plutonium and americium oxides from the human lung. *Health Physics*, **44**, 431–9.

Peirson, D.H., Cambray, R.S., Cawse, P.A., Eakins, J.D. & Pattenden, N.J. (1982) Environmental radioactivity in Cumbria. *Nature*, **300**, 27–31.

Perkins, R.W. & Thomas, C.W. (1980) Worldwide fallout. In: *Transuranic Elements in the Environment*, ed. W.C. Hanson, pp. 53–82. US Dept of Energy, Springfield, Va: NTIS.

Pinder, J.E. III, McLeod, K.W., Simmonds, J.R. & Linsley, G.S. (1985) Normalised specific activities for Pu deposition onto foliage. *Health Physics*, **49**, 1280–3.

Pinder, J.E. III & McLeod, K.W. (1988) Contaminant transport in agroecosystems through retention of soil particles on plant surfaces. *Journal of Environmental Quality*, **17**, 602–7.

Popplewell, D.S., Ham, G.J., Johnson, T.E. & Barry, S.F. (1985) Plutonium in autopsy tissues in Great Britain. *Health Physics*, **49**, 304–9.

Popplewell, D.S., Ham, G.J., McCarthy, W. & Morgan, M. (1989) Isotopic composition of plutonium in human tissue samples determined by mass spectrometry. *Radiation Protection Dosimetry*, **26**, 313–16.

Romney, E.M., Wallace, A., Schulz, R.K., Kinnear, J. & Wood, R.A. (1981) Plant uptake of ^{237}Np, $^{239.240}$Pu, ^{241}Am and ^{244}Cm from soils representing major food production areas of the United States. *Soil Science*, **132**, 40–59.

Ryden, D.J. (1981) Improvements in the detection of airborne plutonium. Report AERE – R 9419. Harwell, Oxfordshire: UKAEA.

Sanders, S.M., Jr & Boni, A.L. (1980) The detection and study of plutonium bearing particles following the re-processing of reactor fuel. In: *Transuranic Elements in the Environment*, ed. W.C. Hanson, pp. 107–44. US Dept of Energy. Springfield, Va: NTIS.

Sherwood, R.J. & Stevens, D.C. (1965) Some observations on the nature and particle size of airborne plutonium in the radiochemical laboratories. *Annals of Occupational Hygiene*, **8**, 93–108.

Shinn, J.H., Essington, E.H., Miller, F.L., O'Farrell, T.P., Orcutt, J.A., Romney, E.M., Shugart, J.W. & Sorom, E.R. (1987) Results of a cleanup and treatment test at the Nevada Test Site: evaluation of vacuum removal of Pu-contaminated soil. *Health Physics*, **57**, 771–9.

Simmonds, J.R., Harrison, N.T. & Linsley, G.S. (1982) *Generalised Derived Limits for Plutonium*. Report DL5. Chilton, Oxon: National Radiological Protection Board.

Simmonds, J.R. & Linsley, G.S. (1982) Parameters for modelling the interception and retention of deposits from atmosphere by grain and leafy vegetables. *Health Physics*, **43**, 679–91.

Stephenson, J., Stevens, D.C. & Morton, D.S. (1971) Removal of radon and thoron daughter products from glass fibre air sample filters. *Annals of Occupational Hygiene*, **14**, 309–19.

Volchok, H.L., Knuth, R. & Kleinman, M.T. (1972) Respirable fraction of plutonium at Rocky Flats. *Health Physics*, **23**, 395–6.

Volchok, H., Schonberg, M. & Toonkel, L. (1977) Pu-239 concentrations in air near Rocky Flats. In: *US Department of Energy Report HASL 315*, pp. 93–109. Springfield, Va: NTIS.

6

○ ○ ○ ○ ○ ○ ○ ○ ○ ○ ○ ○ ○ ○ ○ ○ ○ ○ ○ ○

Mass transfer of radioactive vapours and aerosols

6.1 Application of radioactive measurement techniques

The environmental effects of radioactive aerosols depend on their transfer to surfaces, and this is one reason for studying this process. The other reason is that radioactive aerosols can be used, in the wind tunnel or in the field, to obtain results of general interest. There are several advantages in using radioactivity:

(a) Very sensitive methods are available for measuring deposition on surfaces, and fine detail can be observed, particularly using autoradiography.

(b) By using short-lived activity, the same surfaces can be used repeatedly.

(c) The properties of the fluid or surface are not affected, since the mass of tracer is usually insignificant.

(d) Because the mass deposited is very small – often less than a monolayer – the activity is usually strongly bound to the surface. If there is any re-evaporation or resuspension, this is readily measured.

6.2 Notation

In this connection, it is convenient to use the notation found in the fluid dynamic literature, which is not the same as that normally used for radioactivity. For example C is usually a coefficient, not a concentration.

Notation listed below is used in this chapter.

$\chi(z)$ Concentration of activity per unit volume of air Bq m^{-3}

χ_0 Concentration at surface

χ_1	Concentration at reference height	
A	Activity on surface	$Bq\,m^{-2}$
Q	Flux density	$Bq\,m^{-2}\,s^{-1}$
z	Height above surface	m
z_0, z_v	Roughness length, for momentum, mass transfer	
h	Height of roughness elements	m
$u(z)$	Wind speed at height z	$m\,s^{-1}$
u_1	Wind speed at reference height	$m\,s^{-1}$
u_h	Wind speed at height of roughness elements	$m\,s^{-1}$
u_*	Friction velocity $= (\tau/\rho_a)^{1/2}$	$m\,s^{-1}$
v_g	Velocity of deposition of matter	$m\,s^{-1}$
v_s	Sedimentation velocity	$m\,s^{-1}$
v_t	Velocity of deposition by turbulent transfer	$m\,s^{-1}$
v_m	Velocity of deposition of momentum	$m\,s^{-1}$
v_+	Dimensionless velocity of deposition $= v_g/u_*$	
r	Resistance $= 1/v_g$	$m^{-1}\,s$
r_a, r_b, r_s	Resistances of components of boundary layer	$m^{-1}\,s$
r_+	Dimensionless resistance $= ru_*$	
f_+	Dimensionless resistance to momentum transfer $= u_1/u_*$	
ρ_a, ρ_p	Density of air, particle	$kg\,m^{-3}$
τ	Shearing stress at surface	$N\,m^{-2}$
d_p	Diameter of particle	m
d_A	Aerodynamic diameter $= d_p\rho_p^{1/2}$	
L	Characteristic length	m
D	Molecular diffusivity	$m^2\,s^{-1}$
ν	Kinematic viscosity of air	$m^2\,s^{-1}$
$K(z)$	Eddy diffusivity at height z	$m^2\,s^{-1}$
k	Von Karman's constant, taken equal to 0.4	
C_d	Drag coefficient $= u_*^2/u_1^2$	
C_v	mass exchange coefficient $= v_g/u_1$	
C_i	Impaction efficiency	
C_p	Capture efficiency	
B	Sub-layer transfer coefficient	
Re	Reynolds number $= u_1 L \nu^{-1}$	
Re$_*$	Roughness Reynolds number $= u_* z_0 \nu^{-1}$	
Sc	Schmidt number $= \nu D^{-1}$	
Sh	Sherwood number $= v_g L D^{-1}$	
S_p	Stopping distance of particle	m
Sto	Stokes number $= S_p L^{-1}$	

6.3   Conductances and resistances

The difference $\chi_1 - \chi_0$ in concentration of a vapour or aerosol in the free stream and at a surface is the driving force for deposition. Since the ratio $Q/(\chi_1 - \chi_0)$ has the dimensions of a velocity, it is called the velocity of deposition, denoted v_g. Alternatively, on the electrical analogy, v_g is the conductance and its reciprocal, r is the resistance to mass transfer. If the boundary layer of an airflow over a surface has two or more parts, for example above and below the top of the roughness elements, the resistances of these layers are additive, since

$$r = \frac{\chi_1 - \chi_0}{Q} = \frac{\chi_1 - \chi(h)}{Q} + \frac{\chi(h) - \chi_0}{Q} \qquad (6.1)$$

In the electrical analogy, these are resistances in series. In other instances, it may be the conductances which are additive, as for example where water is transpired from the leaves of plants acting in parallel with evaporation from the ground.

Both v_g and r depend on the reference height chosen for measurement of χ_1, though the effect may not be important when the resistance is concentrated in the boundary layer close to the surface, that is, when $\chi_1 - \chi(h)$ is much less than $\chi(h) - \chi_0$.

Some radioactive vapours are adsorbed or chemisorbed on surfaces so strongly that the boundary condition is χ_0 equal to zero. This is also true of particles in the submicrometric size range. The velocity of gas molecules perpendicular to surfaces is of order 100 m s^{-1}, whereas the resistance of the laminar boundary layer to molecular diffusion usually restricts v_g to a value of the order 0.1 m s^{-1} or less. Hence if the accommodation coefficient, the fraction of molecular collisions which entail sorption at the surface, exceeds about 10^{-3}, the surface will act as a perfect sink.

In the chemical method of investigating transfer through boundary layers, the surfaces are coated with slightly volatile substances such as napthalene, bromobenzene or camphor. The rate of evaporation is measured, and related to the driving force $\chi_0 - \chi_1$, where χ_0 is the concentration corresponding to the vapour pressure of the substance at the surface temperature and χ_1 is the (often negligible) concentration in the free air. In the radioactive method, the adsorbed phase is often less than a monolayer, and the vapour pressure exerted by the deposited vapour is negligible. If, however, the surface is not a perfect sink the resistance χ_0/Q is called the surface resistance, denoted r_s.

For the deposition of particles, bounce-off and blow-off, which must be carefully distinguished, may be considered analogous to the accommodation coefficient and to re-evaporation respectively.

In the analysis of mass transfer, dimensionless numbers are commonly used. Dividing the velocity of deposition by the free stream air velocity gives a transport coefficient C_v sometimes called the Dalton number and analogous to the Stanton number in heat transfer. In discussing mass transfer to or from an object, v_g is made dimensionless by dividing by D/L, where D is the molecular diffusivity of the vapour and L the characteristic length of the object, for example the diameter of a sphere or cylinder or the stream-wise length of a plate. This gives the Sherwood number, Sh. Using the same characteristic length to derive the Reynolds number, the dimensionless numbers are related by

$$Sh = C_v \, Re \, Sc \tag{6.2}$$

where Sc, the Schmidt number, is the ratio of the diffusivity of the vapour to the kinematic viscosity of air.

6.4   Vapour transport to and from surfaces

In early wind tunnel experiments, Chamberlain (1953) liberated elemental ^{131}I vapour and measured deposition on various surfaces. Metal surfaces, and also cellulose-based filter paper, were found to act as perfect sinks for the vapour, deposition being controlled by diffusion through the boundary layer over the surface. In Fig. 6.1, the results for deposition to a flat plate covered with filter paper are compared with the Polhausen equation

$$Sh \, Sc^{-1/3} = 0.66 \, Re^{1/2} \tag{6.3}$$

In calculating Sh and Re, the characteristic length L was the downwind length of the plate and D for I_2 vapour was taken as 8×10^{-6} m^2 s^{-1}. With glass, plastic, and wax-covered surfaces, deposition was much less, and surface resistance to deposition was dominant. The cuticle of the leaves of most plants is waxy. Garland & Cox (1984) compared deposition of ^{132}I$_2$ on bean leaves and on copper replicas chosen for comparison as perfect sinks. The deposition ratio replica/leaf was about four when the relative humidity and light intensity were both high and about 100 when both were low. Garland & Cox also measured transpiration of water from the leaves (Table 3.3). At low RH, transport of both I_2 and water vapour was limited by diffusion through stomatal openings in the leaves. At high RH, there was also sorption of the iodine on the cuticle of the leaf, but this did not correspond to the perfect sink condition.

To investigate mass transfer of a radioactive vapour for which all surfaces are perfect sinks, Chamberlain (1966, 1974) used a vapour of ^{212}Pb (ThB). This was generated by decay of ^{220}Rn (thoron) in a dust-free atmosphere and carried with a flow of filtered air into a wind tunnel (Fig. 6.2). Experiments with real and artificial grass swards in the tunnel are described below. In experiments with bean plants (*Vicia faba*) in the tunnel, the deposition of ^{212}Pb to individual leaves was measured, and analysed in terms of the Sherwood number, using the maximum chord of the leaves as the characteristic length, L.

The wind speed at the level of the top of the bean plants was 2 m s^{-1}, causing fluttering of the leaves, but the Sherwood numbers were only

Fig. 6.1. Transport of gases and particles to and from flat plates and leaves by Brownian diffusion: $\square$, ^{131}I vapour, flat plate (Chamberlain, 1953); $\bigcirc$, ^{212}Pb vapour, bean leaves (Chamberlain, 1974); $\bullet$, water vapour, bean leaves (Grace & Wilson, 1976); +, ×, 0.17-μm particles to pine, oak leaves (Belot, 1975); $\triangle$, $\triangledown$, $\square$, 0.03-μm particles to nettle, beech, white poplar leaves (Little & Wiffen, 1977); line A, theory, laminar flow, $Sh = 0.66Re^{0.5}Sc^{0.33}$; line B, theory, turbulent flow, $Sh = 0.037Re^{0.8}Sc^{0.33}$.

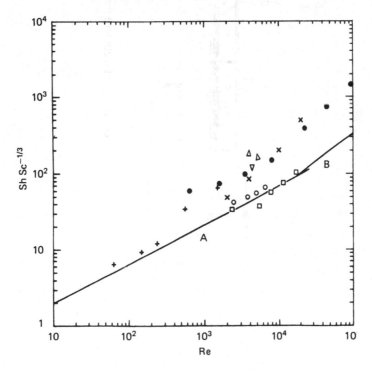

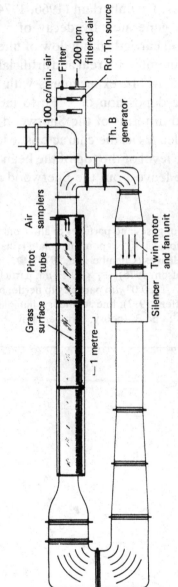

Fig. 6.2. Longitudinal section of tunnel set up for Th.B experiment.

slightly greater than the theoretical value for flat plates aligned to the wind (Fig. 6.1). Also shown are Sherwood numbers for evaporation from fully wetted model leaves (Grace & Wilson, 1976). At high Reynolds numbers, the boundary layer over a flat plate becomes turbulent, and the slope of the relation between Sh and Re becomes steeper.

6.5 Transport of small particles by Brownian diffusion

Equation (6.3) is equivalent to

$$v_g = 0.66\, u^{1/2} v^{-1/6} L^{-1/2} D^{2/3} \tag{6.4}$$

Experiments on transfer of submicrometre radioactive particles to smooth surfaces (Wells & Chamberlain, 1967; Chamberlain et al., 1984) have shown that the dependency of v_g on $D^{2/3}$ holds over many orders of magnitude of D. This means that the transport by Brownian diffusion becomes progressively less effective as the particle size increases. For example a particle of 0.1 μm diameter has a diffusivity of 6.8×10^{-10} $m^2\,s^{-1}$, a factor 1.2×10^4 smaller than that of I_2 vapour. Since D does not depend on the particle density, it is appropriate to discuss transport by Brownian motion in terms of the particle diameter. The aerodynamic diameter, d_A, is equal to $d_p \rho_p^{1/2}$ where ρ_p is the particle density in c.g.s. units (g cm^{-3}) not SI units (kg m^{-3}), and is the appropriate parameter for particles with $d_p > 1\,\mu$m, for which impaction and sedimentation are the mechanisms of deposition.

When petrol containing tetra-ethyl lead is burnt in an engine, particles of inorganic lead are produced in the exhaust. The primary particles are very small, typically 0.01 to 0.05 μm in diameter. The particle size increases with time by aggregation or attachment to larger particles. Little & Wiffen (1978) made radioactive exhaust aerosols by labelling the tetra-ethyl lead with ^{203}Pb, and passed them into the wind tunnel. When the tunnel was operated in the straight-through mode, the median particle size was 0.03 μm, but this was increased by aggregation to 0.2 μm when the tunnel was operated with return flow.

Shoots of nettle, white poplar and beech were placed in the tunnel which was operated at a wind speed of 2.5 m s^{-1}. Beech leaves are smooth, whereas those of white poplar and nettle have hairs. The velocities of deposition to the leaves of the three species are shown in Table 6.1. As expected from the lower diffusivity of the larger particles, v_g was less for the 0.2-μm than for the 0.03-μm particles. Deposition to

Table 6.1. *Deposition of lead aerosols to plant leaves*

Particles median diam. (μm)	D ($m^2 s^{-1}$)	v_g (mm s^{-1})*		
		Nettle	White poplar	Beech
0.03	6.2×10^{-9}	0.60	0.46	0.37
0.4	2.2×10^{-10}	0.15	0.20	0.04

*Velocity of deposition relative to area of both sides of leaf.

the beech leaves was less than deposition to nettle or white poplar, especially for the 0.2-μm particles.

The characteristic lengths of the leaves were 23 mm (nettle), 28 mm (white poplar), and 27 mm (beech). Using these values of L, the wind speed (2.5 m s^{-1}) and the diffusivities as given in Table 6.1, the values of Sh can be calculated and the results are plotted in Fig. 6.1. Also shown are results of Belot (1975), who exposed oak leaves and pine needles to an aerosol of 0.17 μm diameter uranine particles in a wind tunnel, and used the fluorescence of the dye to measure the deposition. The points from Belot's experiments and those of Little & Wiffen lie fairly close to the theoretical line, with the exception of the results of 0.2-μm lead aerosol to nettle and white poplar. These latter show enhanced deposition, probably due to impaction on, or interception by, the hairs on the leaves.

It has been suggested that the planting of trees alongside motorways might reduce the dispersion over from motor exhausts. Little & Wiffen showed that the measured v_g values, when related to the foliar surface of a belt of trees, would not be enough to reduce the amount of lead dispersed significantly.

6.6   Deposition by impaction

When the wind blows past an obstacle, the streamlines of air flow diverge to pass round it. Particles carried in the wind tend to carry straight on and may impact on the obstacle. The efficiency of impaction C_i, is defined as the ratio of the number of impacts to the number of particles which would have passed through the space occupied by the obstacle if it had not been there. If v_g is the velocity of deposition relative to the profile area of the obstacle, then $C_i = v_g/u_1$ where u_1 is the free stream air velocity. C_i is thus analogous to C_d, the drag coefficient of the obstacle.

Aerodynamic theory shows that C_i is a function of the Stokes number defined

$$\text{Sto} = S_p/L \qquad (6.5)$$

S_p is the stopping distance of the particle, which is the distance it would travel in still air if given a velocity u_1. L is the characteristic dimension of the obstacle (the diameter of a cylinder or sphere). For particles in the 1–50 μm range, and u_1 less than 5 m s^{-1}, the approximate stopping distance is given by

$$S_p = v_s u_1 g^{-1} \qquad (6.6)$$

where v_s is the terminal velocity of the particle as given by Stokes Law.

$$v_s = \frac{\rho_p g d_p^2}{18 \, \rho_a \nu} \qquad (6.7)$$

Thus Sto is proportional to the wind speed, the density of the particle and the square of its diameter, and inversely proportional to the size of the obstacle. Both v_s and Sto are functions of d_A, and aerodynamic diameter is used to classify particles deposited by impaction or sedimentation.

Experiments on the impaction of monodisperse water droplets on cylinders (May & Clifford, 1967) have shown C_i correlating with Sto as predicted theoretically. Droplets are almost always captured on impact, but this is not true of solid particles, for which the efficiency of capture C_p may be less than C_i. The same factors which tend to increase C_i, namely large d_p, large u_1 and small L, also tend to increase the possibility of the particle bouncing off the surface, and this may result in a decline in C_p with increase in Sto. Bounce-off of particles from fibres is a well known factor limiting the efficiency of filters.

Paw U (1983) used microphotography to show that 50% of *Lycopodium* spores bounced off leaves of American elm and tulip poplar when the velocity of impact was about 1 m s^{-1}. Little (1977, 1979) used a spinning-top aerosol generator to generate monodisperse polystyrene particles, diameters 2.75, 5 or 8.5 μm, which were labelled with ^{99m}Tc and dispersed in a wind tunnel with wind speeds between 1.5 and 5 m s^{-1}. Figure 6.3 shows the capture efficiency of dry pine needles and needles treated with Vaseline, plotted against Stokes number. Also shown are results of Belot & Gauthier (1975) for the impaction of dye particles on pine shoots and results of Carter (1965) for impaction of ascospore octads on apricot twigs.

The capture efficiency of sticky surfaces for dry particles, and of dry surfaces for moist or sticky particles (ascospores or dye particles), agrees reasonably with theory, but dry surfaces are less efficient for dry particles. Theory and experiments on filter efficiency (Dahneke, 1971; Esmen *et al.*, 1978; Ellenbecker *et al.*, 1980) have shown that the critical parameter is the kinetic energy of the particle. Figure 6.4 shows C_p/C_i, as given by the ratio of catch on non-sticky compared with sticky pine needles (Chamberlain & Little, 1980). Also shown are the results of experiments by Ellenbecker *et al.* (1980) (0.2-μm fly ash particles on 8-μm steel fibres) and Aylor & Ferrandino (1985) (*Lycopodium* particles

Fig. 6.3. Efficiency of impaction of particles on needles and twigs. Data of Belot (1975) using dye particles on pine needles ($\times$), Little (1979) using polystyrene particles on sticky ($\bullet$) and non-sticky ($\bigcirc$) pine needles, and Carter (1965) using ascospore octads on apricot twigs ($\blacktriangle$).

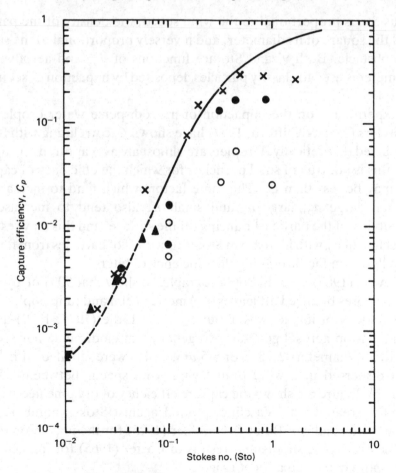

on wheat stalks). Despite the differing particle sizes and types of obstacle, there is a clear correlation between adhesion probability and particle kinetic energy.

A *Lycopodium* spore, or pollen grain of similar size, has an energy of 10^{-4} ergs (10^{-11} J) in a wind speed of 0.75 m s^{-1}, which is typical of conditions near the top of crop canopies or in open woodland. Hence it is likely that pollen grains will bounce off dry leaves or stems, but will be trapped on the stigma of flowers, which are sticky. Spores of plant pathogens, however, are usually themselves moist or sticky, enabling them to adhere to leaves on impact.

6.7   Profiles of wind speed over extended surfaces

Before considering experimental and theoretical data on transfer of vapours and particles to crop canopies, it is necessary to give a highly simplified discussion of boundary layer theory.

The shearing stress, τ, exerted by the wind on the ground entails a downwards flux of momentum. In the aerodynamic boundary layer above the surface, the momentum is transferred by the action of eddy diffusion on the velocity gradient. The friction velocity is defined by $u_*^2 = \tau/\rho_a$ and is a measure of the intensity of the turbulent transfer. Near to a rough surface, the production of turbulance by mechanical forces

Fig. 6.4. Probabilities of particles being captured on impact:
$\triangle$, *Lycopodium* spores and ragweed pollen on wheat stems (Aylor & Ferrandino, 1985); $\blacktriangle$, fly ash particles on steel fibres (Ellenbecker *et al.*, 1980); $\bigcirc$, polystyrene particles on pine needles (Little, 1979).

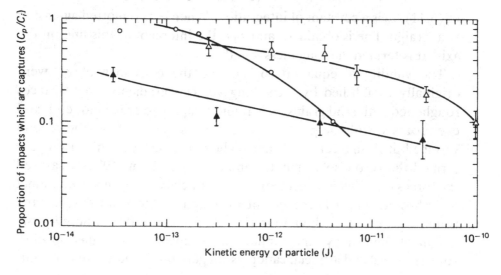

greatly exceeds the production by thermal buoyancy, unless the heat transfer is exceptionally high, and the flow regime is usually adiabatic or nearly so. This being so, the eddy diffusivity at height z is given by

$$K(z) = ku_*z \qquad (6.8)$$

where k is von Karman's constant, taken to be equal to 0.4. To a first approximation, it can be assumed that the eddy diffusivities for momentum, matter and heat are the same.

In fully developed flow over an extensive uniform surface, there is a constant flux layer in which all entities depend only on z. Here the vertical flux of momentum is equal to the stress τ at the surface giving

$$\rho_a K(z) \frac{du}{dz} = \rho_a u_*^2 \qquad (6.9)$$

and, from (6.8)

$$\frac{du}{dz} = \frac{u_*}{kz} \qquad (6.10)$$

giving an integration

$$\frac{u(z)}{u_*} = \frac{1}{k} \log_e \frac{z}{z_0} \qquad (6.11)$$

$$= 5.75 \ln \frac{z}{z_0} \qquad (6.12)$$

In (6.11), z_0 is a constant of integration. When $u(z)$ is plotted against $\ln z$, a straight line is obtained, and z_0 is the intercept of this line on the axis. It is termed the roughness length.

The validity of equation (6.11), and the constancy of z_0, were originally established by measuring velocity gradients over surfaces roughened with sand grains. It is found to apply to grassland, crop and even forests, with the proviso that z is measured not from the ground surface but from a certain distance above the surface. This distance is termed the zero displacement, denoted by d. Thom (1971) measured the forces on individual elements of an artificial canopy in a wind tunnel and showed that d is the height of action of the wind drag on the roughness elements. Thom found $d = 0.76\,h$, $z_0 = 0.09\,h$, where h is the height of the canopy, and similar values have been obtained in most studies of real and artificial canopies. In general, d and z_0 are constants for any particular surface, but both may vary with wind speed if this modifies the crop, for example by flattening it.

Figure 6.5 shows velocity profiles obtained in the wind tunnel depicted in Fig. 6.2. The floor of the tunnel was covered with an artificial sward consisting of plastic spills 75 mm tall and 5 mm wide fixed in a wax substrate (Chamberlain, 1966). With a zero displacement, d, equal to 50 mm, the profile for different wind speeds met on the axis giving $z_0 = 10$ mm, and the slope of the lines gave u_* in accordance with (6.12). When wind speed is normalised by dividing by u_*, the profiles all fall on one line. Figure 6.6 shows u/u_* plotted against $\log(z - d)$ for three surfaces with different z_0. Since the flux of momentum is $\rho_a u_*^2$, and the concentration of momentum at height z is $\rho u(z)$, $u_*^2/u(z)$ is the velocity of deposition of momentum, denoted v_m. Its reciprocal is the resistance to momentum transport and $f_+(z) = u(z)/u_*$ is the normalised resistance.

The profiles in Figs. 6.5 and 6.6 deviate from straight lines when $z - d$ exceeds about 100 mm. The height of the boundary layer, as a pro-

Fig. 6.5. Wind velocity profiles over artificial grass in wind tunnel: O, $u_* - 0.23$, △, $u_* = 0.46$; ▽, $u_* = 0.86$ m s^{-1}.

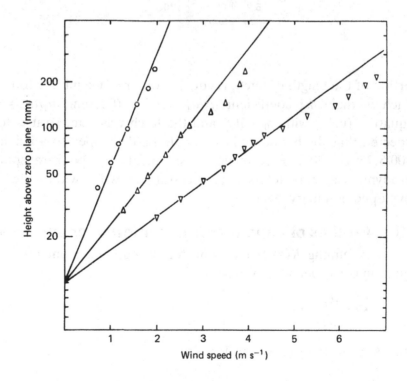

Fig. 6.6. Normalised profiles of wind speed over: $\bigcirc$, artificial grass; $+$, short grass; ∇, towelling: $\triangle$, rough glass.

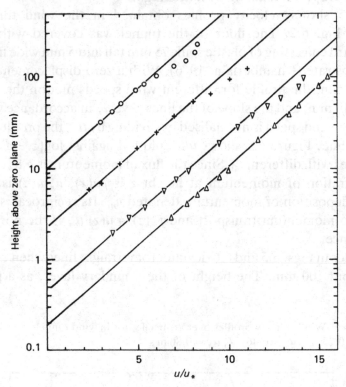

portion of the length of fetch, is equal in order of magnitude to v_m/u_1, which is the drag coefficient u_*^2/u_1^2. At $z = 100$ mm, $u_1/u_* = 5.75$ (equation (6.12) with $z_0 = 10$), and the fetch was 4 m, so it is to be expected that the boundary layer would be developed to a height of $4000/5.75^2$ or 120 mm. To relate measurements of vapour transport to aerodynamical parameters it is necessary to work within the fully developed boundary layer.

6.8   Profiles of vapour concentration over surfaces

Assuming $K(z)$ is the same for vapour and momentum, the equation corresponding to (6.9) is

$$K(z)\frac{\mathrm{d}\chi}{\mathrm{d}z} = Q \qquad\qquad (6.13)$$

With $K(z)$ given by (6.8), this leads to

$$\frac{u_*(\chi(z) - \chi_0)}{Q} = \frac{1}{k} \ln \frac{z}{z_v} \tag{6.14}$$

where z_v is again a constant of integration, termed the roughness length for mass transfer. The processes of absorption of vapour at a surface (or heat transferred from it) are not identical with the processes of absorption of momentum. Vapour transfer to a perfect sink is analogous to skin friction, but there is no analogy to the bluff body effect, whereby pressure forces acting on roughness elements transfer momentum to the surface. Unlike z_0, z_v is not a constant for any one surface, but for surfaces with fibrous roughness elements it only varies slightly with wind speed.

The term on the left of (6.14) is the resistance to mass transfer multiplied by u_*, and is denoted $r_+(z)$. If (6.14) holds, $r_+(z)$ should plot as a straight line against $\ln z$, with the same slope as the plot of $f_+(z)$ against $\ln z$.

Figure 6.7 shows profiles of f_+ and r_+ obtained over artificial grass in the wind tunnel of Fig. 6.2. The f_+ profile is that already shown in Fig. 6.6. Two r_+ profiles are shown. One was obtained with ^{212}Pb vapour. The activity $\chi(z)$ of ^{212}Pb in the tunnel was measured with small suck samplers, and χ_0 was taken as zero. The flux, Q, was deduced from the activity of ^{212}Pb on the artificial grass at the end of the experiment

Fig. 6.7. Dimensionless resistance to transfer of momentum ($\times$), water vapour ($\bigcirc$), and ^{212}Pb vapour ($\bullet$) to artificial grass.

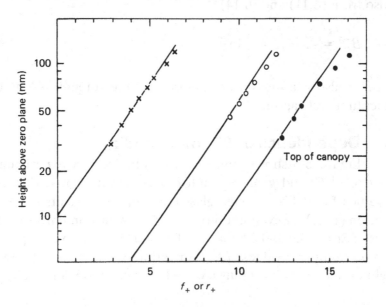

(Chamberlain, 1966). The r_+ profile for water vapour was obtained by measuring the rate of evaporation from the artificial grass, with the spills covered with wetted filter paper. In this case χ_0 was the saturation absolute humidity at the temperature of the spills, $\chi(z)$ was measured, and Q was estimated by periodic weighings of sections of the canopy. The r_+ profiles in Fig. 6.7 are the means of results obtained in a number of experiments at various u_*.

The points in the r_+ profiles diverge from the straight line at somewhat lower values of $z - d$ than the points diverge from the f_+ line. The resistance to vapour transport is greater than the resistance to momentum transport (as shown by the horizontal separation of the profiles), and hence the heights to which the r_+ profiles are fully developed are less.

Owen & Thomson (1963) analysed heat and mass transfer from rough surfaces in terms of a parameter B, termed the sublayer Stanton number, defined by the equation

$$\frac{u(z)}{v_g(z)} = \frac{u(z)}{u_*}\left\{\frac{u(z)}{u_*} + \frac{1}{B}\right\} \tag{6.15}$$

Multiplying both sides by $u_*/u(z)$ this is equivalent to

$$r_+ = f_+ + B^{-1} \tag{6.16}$$

Thus B^{-1} is the displacement parallel to the x-axis of the r_+ relative to the f_+ profiles in Fig. 6.7.

Also from (6.11) and (6.14)

$$B^{-1} = r_+ - f_+ = \frac{1}{k}\ln\frac{z_0}{z_v} \tag{6.17}$$

Thus B^{-1} also is a measure of the ratio of the roughness lengths for momentum and vapour.

6.9   Dependence of B^{-1} on u_* and z_0

Figure 6.8 shows values of B^{-1} obtained in experiments on transfer of ^{212}Pb and water vapour to or from artificial grass. Also shown are values for ^{212}Pb to rough glass together with results of Owen & Thomson (1963) for evaporation of camphor from a similar surface. The values of Sc are 3.2 and 2.8 for camphor and ^{212}Pb vapour respectively. Table 6.2 compares values of B^{-1} for various surfaces with fibrous-type roughness elements, including a value for a bean crop derived by Thom

(1972) by calculating the exchange at individual leaf surfaces. Conclusions from Fig. 6.8 and Table 6.2 are:

(a) B^{-1} for the rough glass surface is much greater than for artificial grass though the roughness length of the rough glass was much less (Fig. 6.6).

(b) The values of B^{-1} for artificial grass show only slight dependence on u_*. Thom (1972) calculated B^{-1} to be proportional to $u_*^{1/3}$.

(c) There is little dependence of B^{-1} on z_0 for surfaces with fibrous roughness elements.

Garratt (1978) analysed data on heat transfer from a heterogeneous surface with high grass and trees ($z_0 = 0.4$ m) and found that B^{-1}

Fig. 6.8. Dependence of B^{-1} on u_*: +, evaporation from artificial grass; ×, ^{212}Pb vapour to artificial grass; △, ^{212}Pb vapour to rough glass; ▲, camphor from rough glass (Owen & Thomson, 1963).

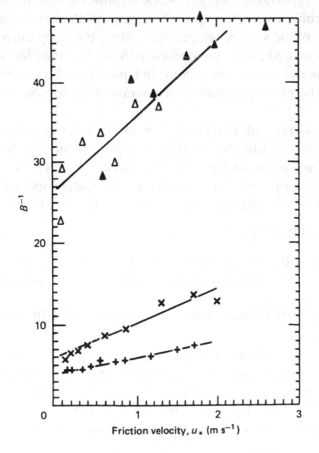

Table 6.2. *Values of* B^{-1} *for transport of* ^{212}Pb *vapour and water vapour at* $u_* = 0.5 \, m \, s^{-1}$

Surface	z_0 (mm)	B^{-1} ^{212}Pb	Water vapour
Artificial grass	10	8.1	5.2
Grass about 70 mm high	8	8.8	
Short grass	2.3	8.8	
Towelling	0.45	7.0	5.5
Bean crop*	70		5.0

*Calculations of Thom (1972).

averaged 6, with little dependence on the roughness Reynolds number $u_* z_0 \nu^{-1}$.

In assessing evapotranspiration, it was sometimes assumed that the temperature profile over a canopy, when extrapolated to $z - d = 0$, gave a measure of the surface temperature. More correctly the profile should be extrapolated to the roughness length for heat transfer, which is nearly the same as z_ν for water vapour. In typical mid-day conditions, this increases the estimated surface temperature by several degrees (Thom, 1972).

When surfaces have bluff roughness elements – for example ploughed soil – skin friction contributes a smaller proportion of the shearing stress, and B^{-1} is increased (Fig. 6.8). Chamberlain *et al.* (1984) found that transfer of ^{212}Pb and ^{131}I to surfaces with widely spaced bluff roughnesses correlated well with the equation (Brutsaert, 1975)

$$B^{-1} = 7.3 \, Re_*^{0.25} \, Sc^{0.5} - 5 \tag{6.18}$$

This result is almost the same as a correlation previously derived by Webb *et al.* (1971) by analysis of heat transfer from rough surfaces.

6.10 Deposition of particles to canopies in the wind tunnel

Wind tunnels provide a partial solution to the study of deposition of particles in the natural environment. The main advantage is that conditions are reproducible. Also, if radioactive aerosols are used, much less activity is needed than in field experiments. The disadvantages are the limitations on the height of the canopy and on the fetch (the length of the canopy upwind of the point of measurement). As a compromise between laboratory and field conditions, movable wind tunnels can be located in the open air, enclosing stretches of natural vegetation (Fig. 6.13, p. 222). At Harwell, wind tunnels have been used to study deposition of radioactivity to natural and artificial canopies and similar experiments have been done in Germany (Möller & Schumann, 1970; Porstendörfer et al., 1980). These have been supplemented by field experiments. In the USA, extensive wind-tunnel and field experiments have been done with particles of dye, using fluorimetric methods to study exchange at surfaces (Sehmel, 1980).

The radioactive aerosols used to measure deposition to surfaces in the Harwell wind tunnel are listed in Table 6.3. Lycopodium spores and ragweed pollen were labelled by making a fluidised bed through which ^{212}Pb (ThB) activity in air was passed. Aitken nuclei, diameter about 0.08 μm, were labelled by attachment of ^{212}Pb while suspended in air. Tri-cresyl phosphate droplets, made in a La Mer generator, and polystyrene spheres, made in a spinning top generator, were labelled by incorporating ^{32}P and ^{51}Cr respectively in the liquids used. All the particles were of approximately unit density.

Figure 6.9 shows the velocity of deposition of particles to grass about 70 mm high (Chamberlain, 1967; Clough, 1975). Results of Porstendörfer et al. (1981) with barley of similar height are also shown. Figure

Table 6.3. *Radioactive aerosols used in wind tunnel*

Vapour or particle	Diameter (μm)	Preparation	Labelling
^{212}Pb	10^{-3}	Generation by decay of ^{220}Rn in dust-free air	^{212}Pb
Aitken nuclei	~0.08	Accumulation of nuclei in laboratory air	^{212}Pb
Droplets of oleic acid	0.5	La Mer generator	
Droplets of tri-cresyl-phosphate	1	La Mer generator	^{32}P
Polystyrene spheres	5	Solution of polystyrene in xylene atomised from spinning top	^{51}Cr
Ragweed pollen	19	Dispersal of particles	^{212}Pb
Lycopodium particles	32	Dispersal of particles	^{212}Pb

6.9 refers to a friction velocity of 0.35 m s^{-1}, corresponding to a wind speed of about 2 m s^{-1} in the tunnel. Measurements by Möller & Schumann (1970) of deposition of radioactive particles in a water/wind tunnel ($u_* = 0.4$ m s^{-1}) are also shown.

The different mechanisms of deposition determine the shape of the curves in Fig. 6.9. For particles of diameter less than 0.2 μm, Brownian diffusion is the dominant mechanism and v_g varies according to $D^{2/3}$, as expected from equation (6.4). The minimum is at d_p between 0.1 and 1 μm, where particles are too large to have appreciable Brownian motion but too small to impact.

The mechanism of deposition in the size range 0.1–1 μm, and the appropriate v_g values, have been the subject of some dispute. Sulphate particles in the urban and suburban atmosphere have median diameters of about 0.5 μm (Heard & Wiffin, 1969; Whitby, 1978). Using the results of Fig. 6.9, and weighting v_g according to the mass of sulphate in various size ranges, Garland (1978) calculated a mean value of v_g for sulphate aerosol of 1.0 mm s^{-1}. Nicholson & Davies (1987) measured the profile of SO_4^{2-} concentrations, and also wind speed, above agricul-

Fig. 6.9. Deposition of radioactive particles to surfaces in wind tunnels ($u_* = 0.35$ mm s^{-1}): $\bigcirc$, grass (Chamberlain, 1967); $\triangle$, grass (Clough, 1975); $\triangledown$, barley (Porstendörfer *et al.*, 1981); – – –, water surface (Möller & Schumann, 1970).

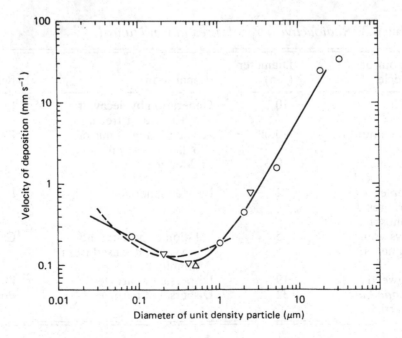

tural land in East Anglia, and deduced a mean v_g of $0.7\ \mathrm{mm\ s^{-1}}$. The low deposition velocity of sulphate particles and other photochemical products of pollution explain why photochemical smog is so persistent during fine weather.

Sedimentation of particles contributes significantly to deposition when the diameter exceeds about $5\ \mu\mathrm{m}$. Instead of (6.13), the flux is given by

$$Q = v_s \chi(z) + K(z) \frac{\mathrm{d}\chi(z)}{\mathrm{d}z} \tag{6.19}$$

or, dividing by $\chi(z)$

$$v_g = v_s + v_t \tag{6.20}$$

when v_t is the turbulent transfer velocity of matter.

Integration of (6.19) leads to

$$v_g = \frac{v_s}{1 - (z_v/z_1)^\alpha} \tag{6.21}$$

where z_1 is the reference height, z_v is the height of the effective sink for particles, and $\alpha = v_s/ku_*$.

Figure 6.10 shows the velocity of deposition of *Lycopodium* spores to grass and to sticky artificial grass in the wind tunnel (Chamberlain, 1967). In these experiments, the plastic strips of the artificial grass were treated with Vaseline. This had the effect of cushioning the impact of the spores and preventing bounce-off or blow-off. The terminal velocity of the spores was found to be $19\ \mathrm{mm\ s^{-1}}$ (the spores have a maximum diameter of $32\ \mu\mathrm{m}$ but they are neither spherical nor smooth, so Stokes Law does not strictly apply). Curve A in Fig. 6.10 is equation (6.21) with z_v equated to z_0 (10 mm). Curve B is (6.20) with v_t equated to v_m, the velocity of deposition of momentum at the reference height ($z - d = 75$ mm). The difference between the curves is small but significant when v_s is not very small in comparison with u_*. A profile of χ can be deduced from (6.19), and when α exceeds about 0.1 a plot of χ against $\ln(z - d)$ is not linear as in Fig. 6.5, but is concave towards the z-axis.

The experimental points for the sticky artificial grass in Fig. 6.10 lie close to the theoretical curves, showing that the *Lycopodium* particles were deposited at the maximum rate allowed by the aerodynamic resistance. Deposition to real grass was limited by bounce-off. For both surfaces v_t was proportional to u_*, the normalised turbulent deposition velocity v_+ being 0.18 and 0.04 for sticky and real grass respectively.

Figure 6.11 shows the velocity of deposition of *Lycopodium* spores to bare soil in the wind tunnel (Chamberlain & Chadwick, 1972). The roughness length was 3 mm. When the soil was wet, v_t increased with u_*, but levelled off at high speeds (in the field, $u_* = 1$ m s^{-1} would correspond to $u = 20$ m s^{-1} at $z = 10$ m). With dry soil, v_g hardly

Fig. 6.10. Deposition of *Lycopodium* spores to grass ($\triangle$) and sticky artificial grass ($\bigcirc$). Lines A, B, are from equations (6.21) and (6.20), respectively.

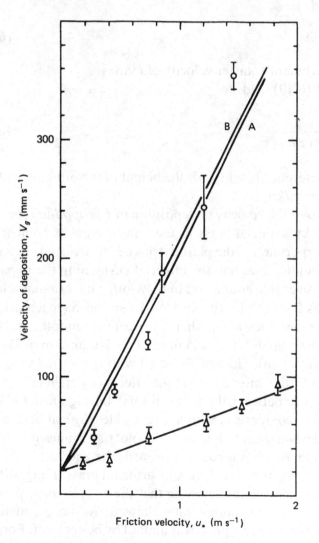

increased with u_*, and v_t was small. At $u_* = 1$ m s^{-1}, saltation of soil particles began.

The deposition velocities to grass of particles with diameters between 1 and 5 μm were also found to be proportional to u_* (Chamberlain, 1967), but for these particles impaction has much less effect. Particles smaller than about 5 μm diameter are unlikely to bounce off surface in normal ambient conditions. The presence of micro-roughness elements on surfaces increases v_g for these particles, and Chamberlain (1967) found more deposition to real leaves than to smooth sticky artificial leaves when the particle size was less than 5 μm.

6.11 Field experiments on deposition of particles

In a wind tunnel, the direction and velocity of the airflow are predetermined, but in the field they are variable. This presents problems in the sampling of airborne particles, as discussed by May *et al.* (1976). If suck samples are used, the flow should be isokinetic, which requires that the sampler be aligned upwind and that the flow velocity into the sampler equals the wind speed. If these conditions are not met

Fig. 6.11. Deposition of *Lycopodium* spores to soil: A, dry soil; B, wet soil; C, theoretical values assuming surface is a perfect sink.

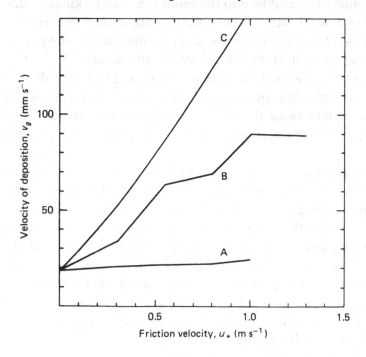

the streamlines will bend as the air enters the sampler and particles which deviate from the streamline will not be sampled correctly. The inaccuracy becomes serious when the particle size exceeds 10 μm.

Plant pathologists need to study the dispersion of spores in the field. To avoid the need for a power supply, and the difficulties of arranging isokinetic samplers, passive samplers are often used (Gregory, 1973). Horizontal microscope slides are used to collect spores by sedimentation and vertical cylinders are used to collect them by impaction. The efficiency of captures by the cylinders depends on the wind speed, and requires calibration in the wind tunnel.

Field experiments with *Lycopodium* spores were carried out at Harwell and at Rothamsted (Chamberlain, 1967; Chamberlain & Chadwick, 1972). The spores were made radioactive by absorbing ^{131}I onto them (about 10^8 Bq per g of spores) and dispersed at a height of 0.5 m above ground. At distances of 20–50 m downwind, samplers were arranged to measure the airborne concentration and the deposition on the crop was measured. Vertical sticky cylinders, 6.5 mm in diameter, were used. Their efficiency was measured in the wind tunnel and found to agree with the theoretical predictions. Also, a sampling arrangement using small portable wind tunnels, as described by May *et al.* (1976), was used. A probe was arranged to sample isokinetically relative to the flow in the tunnel. The airflow into the tunnel was not isokinetic relative to the ambient wind, but the curvature of the streamlines entering the tunnel was large compared to the stopping distances of the particles, so particles entering the tunnel followed the streamlines.

The results of the field experiments are summarised in Table 6.4. The velocity of deposition of *Lycopodium* spores to dry grass and cereal crops was about twice the sedimentation velocity, so v_s and v_t were about equal. When the cereal crop was thoroughly wetted, v_t was about twice v_s.

Oceult precipitation is the capture of wind-driven fog or cloud droplets by vegetation, under conditions where minimal precipitation is found in rain gauges. Dollard & Unsworth (1983) and Gallagher *et al.*, (1988) measured the gradients of fog-water concentration and of wind speed over grassland at Great Dun Fell in the Pennines and obtained results shown in the last row of Table 6.4. Dollard & Unsworth estimated that occult precipitation to grassland would give the equivalent of 0.14 mm rainfall during a foggy period lasting 8 h. Occult precipitation is more significant in mountainous regions, especially if forested, as considered below.

Table 6.4. *Field experiments on deposition of particles*

	Surface	No. of exps.	u_* (mean) (ms^{-1})	Velocity of deposition (mm s^{-1})				Reference
				v_g (mean ± S.E.)	v_s	v_t	v_m	
Dry Surfaces								
Lycopodium	Grass	7	0.48	37 ± 7	19	18	72	Chamberlain (1967)
Lycopodium	Cereals	5	0.54	36 ± 4	19	17	89	Chamberlain & Chadwick (1972)
Wet Surfaces								
Lycopodium	Cereals	4	0.39	59 ± 11	19	40	65	Chamberlain & Chadwick (1972)
Fog droplets	Grass	10	0.31	45 ± 5	18	27	33	Dollard & Unsworth (1983)
Fog droplets	Grass	9	0.56	60 ± 3	9	51	52	Gallagher *et al.* (1988)

In the last two columns of Table 6.4 the turbulent deposition velocities of mass and momentum are shown. For droplets, v_t and v_m are
approximately equal, indicating that capture by the surface is efficient,
and aerodynamic resistance the limiting factor.

6.12 Deposition of particles to forests

The capture of acid particles, or acid droplets, by forests is
considered an important cause of acidification of soils and water courses
in mountain regions (Lovett, 1984; Lovett & Reiners, 1986). Field
experiments on the required scale are hardly feasible. Several authors
have made calculations of the wind profile within the forest, and the
capture efficiency of model leaves and twigs in the canopy. Figure 6.12
shows the results of calculations by Belot (1975), Slinn (1982), and
Lovett (1984). When expressed in terms of the normalised velocity of
deposition v , there is little difference in Fig. 6.9 between the calculated

Fig. 6.12. Normalised velocity of deposition: A, radioactive particles to
grass, $u_* = 0.35$ m s^{-1}; B, C, D, calculations for forests by Slinn
(*Eucalyptus*, $u_* = 0.7$ m s^{-1}), Belot (pine, $u_* = 0.6$ m s^{-1}) and Lovett (fir,
$u_* = 1.1$ m s^{-1}). Reference height = twice canopy height.

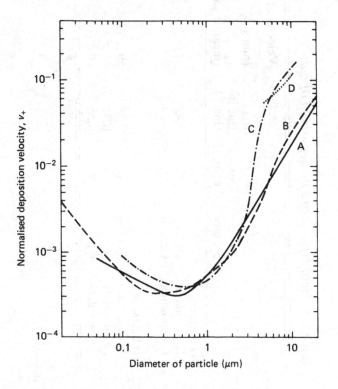

results and the experimental results, over the range of d_p from 0.05 to 2 μm. The roughness lengths of the model forests were 15–20 times greater than the roughness length of the grass swards in the wind tunnel experiments, so it appears that v_+ is not very dependent on z_0 for particles in this size range.

For coniferous forests, the calculated v_+ increases rapidly as droplet diameter increases to 10 μm. Also, u_* is typically several times greater over a forest than over grassland, so the disparity is greater in terms of v_t. Lovett & Reiners (1986) found v_t of cloud droplets to a subalpine fir forest to be 300 mm s^{-1}, increasing possibly to 2000 mm s^{-1} on the lee side of gaps in the canopy. In these conditions, occult precipitation is the equivalent of 0.1 to 0.3 mm h^{-1} of rainfall (Lovett, 1984). Much of the intercepted water re-evaporates, but ions dissolved in the droplets remain on the leaves and are potentially damaging.

6.13   Resuspension of particles

Most radioactive particles and vapours, once deposited, are held rather firmly on surfaces, but resuspension does occur. A radioactive particle may be blown off the surface, or, more probably, the fragment of soil or vegetation to which it is attached may become airborne. This occurs most readily where soils and vegetation are dry and friable. Most nuclear bomb tests and experimental dispersions of fissile material have taken place in arid regions, but there is also the possibility of resuspension from agricultural and urban land, as an aftermath of accidental dispersion. This is particularly relevant to plutonium and other actinide elements, which are very toxic, and are absorbed slowly from the lung, but are poorly absorbed from the digestive tract. Inhalation of resuspended activity may be the most important route of human uptake for actinide elements, whereas entry into food chains is critical for fission products such as strontium and caesium.

Classic work on resuspension was done by Bagnold (1954), including field experiments in the Western Desert of Egypt and experiments in a wind tunnel. Bagnold studied resuspension of sand particles, typically 50–200 μm in diameter. For such particles, the sedimentation velocity is comparable with or larger than the vertical eddy wind velocities near the ground, which are of order u_*. Movement occurs by saltation. Particles strike the ground and rebound, or dislodge other particles, and proceed in a series of hops. In desert sands, there are few small particles, because any such have been winnowed away, and nearly all movement takes place within a metre or so of the ground. In these conditions, Bagnold found that the horizontal flow of sand was proportional to u_*^3.

Saltation can occur in agricultural regions when strong winds erode dry soil, but resuspension occurs much more generally. Particles in the 5–30 μm size range become airborne in various ways and are transported upwards by eddies and thermal air currents.

Shinn et al. (1976) measured the concentration of dust in air, at various heights above ground, at the Nevada Test Site and at a site in Texas. Their observations, and others which they found in the literature, fitted a power law

$$\chi(z) = \chi_1 z^{-p} \tag{6.22}$$

where χ_1 is the concentration at the reference height (1 m) and p has values between 0.25 and 0.35. They deduced that the vertical turbulent flux of dust at the reference height was

$$Q = ku_* z_1 \frac{\mathrm{d}\chi}{\mathrm{d}z} = ku_* p \chi_1 \tag{6.23}$$

At the Nevada Test Site, where saltation was not evident, Shinn et al. found a correlation

$$\chi_1 = 6.1 \, u_*^{2.09} \tag{6.24}$$

where χ_1 is in mg m^{-3} and u_* is in m s^{-1}.

Hence, taking $p = 0.3$ and $k = 0.4$, they deduced

$$Q = 0.12 \, u_* \chi_1 \tag{6.25}$$

$$= 0.73 \, u_*^{3.09} \tag{6.26}$$

where Q is in mg m^{-2} s^{-1}.

The theoretical concentration in air downwind of an area source as a function of the roughness length and atmospheric stability has been given by Wilson (1982). Shinn et al. found $z_0 = 0.02$ m at the Nevada site. The dimensionless concentration $u_* \chi_1 / Q$ from (6.25) is 8.3, and this would agree with Wilson's calculations for average (neutral) stability if the fetch of wind over the source of resuspended dust was 10 km.

From (6.26) the dependence of the vertical flux of dust on u_* at the Nevada Test Site was the same as the dependence of horizontal advection on u_* in Bagnold's work. At the Texas site, where severe soil erosion, with saltation, occurred in strong winds, χ_1 and Q varied much more strongly with u_*.

The resuspension of radioactive fallout can be related to the activity originally deposited. If A (Bq m^{-2}) is the activity on the ground, χ_1 (Bq m^{-3}) the airborne activity, and Q (Bq m^{-2} s^{-1}) the vertical flux, the resuspension factor is defined.

$$K_r = \chi_1/A \qquad (6.27)$$

and the resuspension rate

$$\Lambda = Q/A \qquad (6.28)$$

The units of K_r and Λ are m^{-1} and s^{-1} respectively, so K_r/Λ, equal to χ_1/Q, has the dimensions of a resistance.

Stewart (1966) has summarised values of K_r found by monitoring airborne activity a few days after fallout from the bomb tests at Maralinga and Monte Bello Island. A range of values from 10^{-4} to 10^{-7} m^{-1} was obtained, but where the ground was disturbed by vehicles, K_r reached 7×10^{-4} m^{-1}.

Anspaugh et al. (1975, 1976) studied resuspension at the GMX location on the Nevada Test Site, where plutonium was disseminated by small non-nuclear explosions about 30 years ago. To obtain representative values of the resuspension factor, ideally a large and uniform area of deposited activity is required, in order that there should be a constant flux layer in the air near the ground. The fallout of Pu at the GMX site is non-uniform, so Anspaugh et al. analysed their measurements of χ in relation to a model calculation of the concentration expected from the areal source.

The values of Λ derived in this way correlated well with u_*^3, the ratio Λ/u_*^3 having a median value of 1×10^{-10} s^2 m^{-3}. The range of Λ was from 3×10^{-12} to 5×10^{-10} s^{-1} and the range of K_r from 9×10^{-11} to 5×10^{-9} m^{-1}. Most of the Pu was associated with particles of diameter less than $10\,\mu$m.

A mean value of 3×10^{-10} m^{-1} for K_r was deduced by Shinn et al. (1983) from measurements of airborne Pu over a bare field near the Savannah River Processing Plant. Here the activity median aerodynamic diameter was $3\,\mu$m. The main reason why K_r measured in Nevada and North Carolina was much lower than K_r measured at Maralinga and Monte Bello is the effect of ageing on the characteristics of the deposited material.

Although resuspension of soil is usually considered a problem of arid climates, it also contributes to air pollution in temperate regions. Some of the socalled soil-derived elements (for example, aluminium, scandium) in air near the ground derive from resuspension, but it is difficult with stable elements to discriminate this source from emissions of fly ash from power stations, since fly ash has similar elemental composition to soil.

Hötzl *et al.* (1989) measured the airborne concentration of ^{137}Cs in Germany in the year subsequent to the Chernobyl accident. Though low compared to the concentrations immediately after the accident, the levels were higher than could be ascribed to the lingering effects of weapon tests, and were found to correlate, in different locations, with the amount of Chernobyl fallout. Comparing the airborne concentration with that of ^{137}Cs in the top 10 mm of soil, Hötzl *et al.* deduced a resuspension factor $K_r = (3 \pm 1) \times 10^{-9}$ m^{-1}. Concentrations were found to increase with wind speed as $u^{1.8}$, a result very similar to that found in Nevada (equation (6.24)).

Resuspension of dust from the floor is an important source of indoor pollution. Jones & Pond (1966) studied resuspension of plutonium from the floor of a laboratory which had been deliberately contaminated. Droplets were dispersed over the floor and allowed to dry out. Pu was present in the droplets either as a suspension of particles, mass median diameter 15 μm, or as nitrate in solution. The activity per unit area was measured with a floor probe. After 16 h had elapsed, airborne Pu was measured while operators performed various tasks in the laboratory.

The resuspension factor K_r was found to depend on the amount of movement in the laboratory, with values ranging from 10^{-5} to 10^{-4} m^{-1} (oxide particles) and from 10^{-6} to 10^{-5} m^{-1} (Pu applied in solution).

Fig. 6.13. Diagram of wind tunnel used to study resuspension.

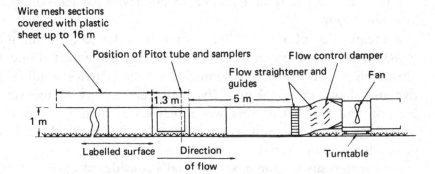

6.14   Resuspension – wind-tunnel experiments

Garland (1979, 1982, 1983) used the wind tunnel shown in Fig. 6.13 to measure resuspension of radioactive particles from grassland at Harwell. The fan and motor were mounted on a turntable, and the working section could be positioned as required. Radioactive particles were deposited on a strip of grass about 10 m long, and air was then drawn over it in the tunnel. Samplers measured the amount of resuspended activity in the air downwind of the strip. The horizontal flux of activity was deduced and expressed as the rate constant Λ of resuspension.

The following tracers were used:

(a)   Monodisperse 5-μm diameter spheres of ferric oxide, labelled with ^{59}Fe, applied as a spray.

Fig. 6.14. Rate of resuspension from grass after exposure to wind for 10 h of: ×, 5-μm particles; ●, 2-μm particles; ○, FeCl$_3$ in solution; +, sub-micrometre particles.

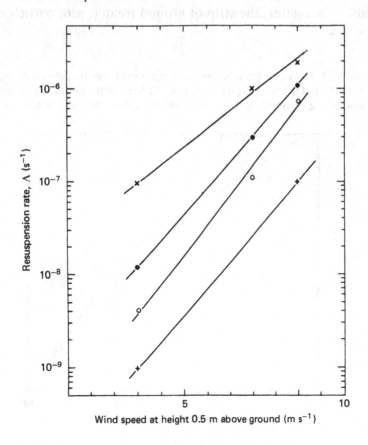

(b) Similar 2-μm diameter spheres.
(c) FeCl$_3$, labelled with ^{59}Fe, applied in solution, and allowed to dry.
(d) Tungstic oxide particles, of submicrometric size, irradiated to form ^{185}W, applied by settling from air over the strip.

Figure 6.14 shows the resuspension rates after 10-h runs at various wind speeds. Resuspension is clearly related to particle size. The activity applied as FeCl$_3$ in solution was probably resuspended while attached to particles of leaf cuticle. Garland did not give the value of u_*, but it was probably about 1/10th of the wind speed at height 0.5 m. Values of Λ/u_*^2 ranged from about 5×10^{-8} s^2 m^{-3} for the submicrometre particles to 3×10^{-6} s^2 m^{-3} for the 5-μm particles, and were several orders of magnitude greater than the value 1×10^{-10} s^2 m^{-3} deduced by Anspaugh et al. (1975) from the Nevada work.

Figure 6.15 shows the effect of ageing on the resuspension of the submicrometre particles. The wind tunnel was operated continuously for 100 h with a wind speed of 12 m s^{-1}, and periodic measurements of Λ were made. Thereafter, the strip of ground treated with particles was

Fig. 6.15. Resuspension of submicrometre particles of tungstic oxide: up to 100 h – wind tunnel operated at 12 m s^{-1}; after 100 h – periods of natural weathering (hatched) in intervals between further operation of wind tunnel.

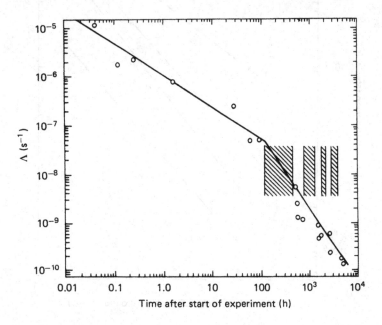

allowed to weather naturally in the open air, in the intervals between successive further periods of operation.

During the first 100 h, Λ was approximately inversely proportional to the elapsed time, and the same was true in other experiments with different tracers. After 250 days, Λ was about 10^{-10} s^{-1}, within the range of values found by Anspaugh *et al.* (1975) at the Nevada Test Site.

References

Anspaugh, L.R., Shinn, J.H., Phelps, P.L. & Kennedy, N.C. (1975) Re-suspension and re-distribution of plutonium in soils. *Health Physics*, **29**, 571–82.

Anspaugh, L.R., Phelps, P.L., Kennedy, N.C., Shinn, J.H. & Reichmann, J.M. (1976). Experimental studies on the re-suspension of plutonium from aged sources at the Nevada Test Site. In: *Atmospheric-Surface Exchange of Particulate and Gaseous Pollutants*, ed. R.J. Engelmann & G.A. Sehmel. CONF 740921. NTIS. Springfield, Va.

Aylor, D.E. & Ferrandino, F.J. (1985) Rebound of pollen and spores during deposition on cylinders by inertial impaction. *Atmospheric Environment*, **19**, 803–6.

Bagnold, R.A. (1954) *The Physics of Blown Sand and Desert Dunes*. London: Methuen.

Belot, Y. (1975) Etude de la captation des pollutants atmosphériques par les végétaux. Commissariat a l'Energie Atomique, Fontenay-aux-Roses, Paris.

Belot, Y. & Gauthier, D. (1975) Transport of micronic particles from atmosphere to foliar surfaces. In: *Heat & Mass Transfer in the Biosphere*, ed. D.A. de Vries & N.H. Algan, pp. 583–91. Washington D.C.: Scripta Book Co.

Brutsaert, W. (1975) A theory for local evaporation (or heat transfer) from rough or smooth surfaces at ground level. *Water Resources Research*, **11**, 543–50.

Carter, M.V. (1965) Ascospore deposition in *Eutypa armeniacae*. *Australian Journal of Agricultural Science*, **16**, 825–36.

Chamberlain, A.C. (1953) Experiments on the deposition of iodine-131 vapour onto surfaces from an airstream. *Philosophical Magazine*, **44**, 1145–53.

(1966) Transport of gases to and from grass and grass-like surfaces. *Proceedings of the Royal Society*, A, **290**, 236–65.

(1967) Transport of *Lycopodium* spores and other small particles to rough surfaces. *Proceedings of the Royal Society*, A, **296**, 45–70.

(1974) Mass transfer to bean leaves. *Boundary-layer Meteorology*, **6**, 477–86.

Chamberlain, A.C. & Chadwick, R.C. (1972) Deposition of spores and other particles on vegetation and soil. *Annals of Applied Biology*, **71**, 141–58.

Chamberlain, A.C. & Little, P. (1980) Transport and capture of particles by vegetation. In: *Plants and their Atmospheric Environment*, ed. J. Grace, E.D. Ford & P. Jarvis, pp. 147–73. Blackwell: Oxford.

Chamberlain, A.C., Garland, J.A. & Wells, A.C. (1984) Transport of gases and particles to surfaces with widely spaced roughness elements. *Boundary-layer Meteorology*, **29**, 343–60.

Clough, W.S. (1975) The deposition of particles on moss and grass surfaces. *Atmospheric Environment*, **9**, 1113–19.

Dahneke, B. (1971) The capture of aerosol particles by surfaces. *Journal of Colloid & Interface Science*, **37**, 342–53.

Dollard, G.J. & Unsworth, M.H. (1983) Field measurements of turbulent fluxes of wind driven fog drops to a grass surface. *Atmospheric Environment*, **17**, 775–80.

Ellenbecker, M.J., Leith, D. & Price, J.M. (1980) Impaction and bounce at high Stokes number. *Journal of the Air Pollution Control Association*, **30**, 1224–7.

Esmen, N.A., Ziegler, P. & Whitfield, R. (1978) The adhesion of particles upon impaction. *Journal of Aerosol Science*, **9**, 547–66.

Gallagher, M.W., Choularton, T.W., Morse, A.P. & Fowler, D. (1988) Measurements of the size dependence of cloud droplet deposition at a hill site. *Quarterly Journal of the Royal Meteorological Society*, **114**, 1291–303.

Garland, J.A. (1978) Dry and wet removal of sulphur from the atmosphere. *Atmospheric Environment*, **12**, 349–62.

(1979) Resuspension of particulate matter from grass and soil. AERE Report – R9452, London HMSO.

(1982) Resuspension of particulate material from grass: experimental programme 1979–1980. AERE Report – R 10106, London HMSO.

(1983) Some recent studies of the resuspension of deposited material from soil and grass. In: *Precipitation Scavenging, Dry Deposition and Resuspension*, ed. H.R. Pruppacher, R.G. Simonin & W.G.N. Slinn, pp. 1087–97. Amsterdam: Elsevier.

Garland, J.A. & Cox, L.C. (1984) The uptake of elemental iodine vapour by bean leaves. *Atmospheric Environment*, **18**, 199–204.

Garratt, J.R. (1978) Transfer characteristics for a heterogeneous surface of large aerodynamic roughness. *Quarterly Journal of the Royal Meteorological Society*, **104**, 491–502.

Grace, J. & Wilson, J. (1976) The boundary layer over a *Populus* leaf. *Journal of Experimental Botany*, **27**, 231–41.

Gregory, P.H. (1973) *The Microbiology of the Atmosphere*. 2nd edn. London: Leonard Hill.

Heard, M.J. & Wiffen, R.D. (1969) Electron microscopy of natural aerosols and the identification of particulate ammonium sulphate. *Atmospheric Environment*, **3**, 337–40.

Hötzl, H., Rosner, G. & Winkler, R. (1989). Long term behaviour of Chernobyl fallout in air and precipitation. *Journal of Environmental Radioactivity*, **10**, 157–71.

Jones, I.S. & Pond, S.F. (1966) Some experiments to determine the re-suspension factor of plutonium from various surfaces. In: *Surface Contamination*, ed. B.R. Fish, pp. 83–92. Oxford: Pergamon.

Little, P. (1977) Deposition of 2.75, 5.0 and 8.5 μm particles to plant and soil surfaces. *Environmental Pollution*, **12**, 293–305.

(1979) Unpublished work (see Chamberlain & Little, 1980).

Little, P. & Wiffen, R.D. (1977) Emission and deposition of petrol engine exhaust Pb. I. Deposition of exhaust Pb to plant and soil surfaces. *Atmospheric Environment*, **11**, 437–47.

Lovett, G.M. (1984) Rates and mechanisms of cloud water deposition to a sub-alpine balsam fir forest. *Atmospheric Environment*, **18**, 361–77.

Lovett, G.M. & Reiners, W.A. (1986) Canopy structure and cloud water deposition in sub-alpine coniferous forests. *Tellus*, **38B**, 319–27.

May, K.R. & Clifford, R. (1967) The impaction of aerosol particles as cylinders, spheres, ribbons and discs. *Annals of Occupational Hygiene*, **10**, 83–95.

May, K.R., Pomeroy, N.P. & Hibbs, N. (1976) Sampling techniques for large windborne particles. *Journal of Aerosol Science*, **7**, 53–62.

Möller, V. & Schumann, G. (1970) Mechanisms of transport from the atmosphere to the earth's surface. *Journal of Geophysical Research*, **75**, 3013–19.

Nicholson, K.W. & Davies, T.D. (1987) Field measurements of the dry deposition of particulate sulphate. *Atmospheric Environment*, **21**, 1561–71.

Owen, P.R. & Thomson, W.R. (1963) Heat transfer across rough surfaces. *Journal of Fluid Mechanics*, **15**, 321–34.

Paw U, K.T. (1983) The rebound of particles from natural surfaces. *Journal of Colloid & Interface Science*, **93**, 442–52.

Porstendörfer, J., Robig, G. & Ahmed, A. (1980) Washout and dry deposition of atmospheric aerosols. In: *Radiation Protection*, vol. I, pp. 583–7. Oxford: Pergamon.

Sehmel, G.A. (1980) Particle and gas dry deposition: a review. *Atmospheric Environment*, **14**, 983–1011.

Shinn, J.H., Kennedy, N.C., Koral, J.S., Clegg, B.R. & Porch, W.M. (1976) Observations of dust flux in the surface boundary layer for steady and non-steady cases. In: *Atmosphere–Surface Exchange of Particulate and Gaseous Pollutants*, ed. R.J. Engelmann & G.A. Sehmel, CONF 740921. Springfield Va.: NTIS.

Shinn, J.H., Homan, D.N. & Gay, D.D. (1983) Plutonium aerosol fluxes and pulmonary exposure rates during re-suspension from bare soils near a chemical separation facility. In: *Precipitation Scavenging, Dry Deposition and Re-suspension*, ed. H.R. Pruppacher, R.G. Semonin & W.G.N. Slinn, pp. 1131–43. Amsterdam: Elsevier.

Slinn, W.G.N. (1982) Predictions for particle deposition to vegetative canopies. *Atmospheric Environment*, **16**, 1785–94.

Stewart, K. (1966) The resuspension of particulate material from surfaces. In: *Surface Contamination*, ed. B.R. Fish, pp. 63–74. Oxford: Pergamon.

Thom, A.S. (1971) Momentum absorption by vegetation. *Quarterly Journal of the Royal Meteorological Society*, **97**, 414–28.

(1972) Momentum, mass and heat exchange of vegetation. *Quarterly Journal of the Royal Meteorological Society*, **98**, 124–34.

Webb, R.L., Eckert, E.R.G. & Goldstein, R.J. (1971) Heat transfer and friction in tubes with repeated-rib roughness. *International Journal of Heat & Mass Transfer*, **14**, 601–17.

Wells, A.C. & Chamberlain, A.C. (1967) Transport of small particles to vertical surfaces. *British Journal of Applied Physics*, **18**, 1793–9.

Whitby, K.T. (1978) The physical characteristics of sulfur aerosols. *Atmospheric Environment*, **12**, 135–59.

Wilson, J.D. (1982) Turbulent dispersion in the atmospheric surface layer. *Boundary-Layer Meteorology*, **22**, 399–420.

7

○ ○ ○ ○ ○ ○ ○ ○ ○ ○ ○ ○ ○ ○ ○ ○ ○ ○

Studies with radioactive particles and human subjects

7.1 Introduction

Concern about possible inhalation hazards was felt at an early stage in the development of the nuclear industry. Experiments were done with animals to study deposition of radioactivity in the lung, entry into the bloodstream and transfer to organs of concentration. Information has also been obtained from cases of accidental human exposure. The Task Group on Lung Dynamics (1966) reviewed these data and formulated recommendations on which are based the permissible levels of airborne activity.

No attempt is made here to review the basis of these recommendations, but some human studies with radioactive aerosols are described, with particular reference to those relevant to exposure to atmospheric pollutants, such as tobacco smoke and lead aerosols.

The effects of inhaled particles depend on the deposition in the upper airways and in the lung.

The Task Group on Lung Dynamics (1966) considered the respiratory tract as consisting of three compartments (Fig. 7.1):

  (i) Naso-pharyngeal (NP) region, or upper respiratory tract. This comprises the nose, mouth, naso-pharynx and oro-pharynx.

 (ii) Tracheo-bronchial (TB) region. This begins at the larynx, and comprises the trachea, bronchi and bronchioles. The branching airways are smaller and more numerous at each division, ending with the terminal bronchioles which are about 0.5 mm in diameter.

(iii) Pulmonary (P) or alveolar region. The respiratory bronchioles and alveoli perform the function of the lung in exchanging oxygen and CO_2.

228

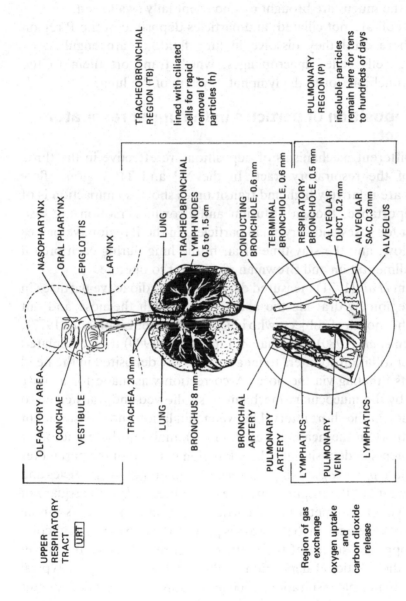

UPPER RESPIRATORY TRACT URT

OLFACTORY AREA
NASOPHARYNX
CONCHAE
ORAL PHARYNX
VESTIBULE
EPIGLOTTIS
LARYNX

TRACHEA, 20 mm
LUNG
TRACHEO-BRONCHIAL LYMPH NODES 0.5 to 1.5 cm
LUNG
BRONCHUS 8 mm
CONDUCTING BRONCHIOLE, 0.6 mm
TERMINAL BRONCHIOLE, 0.6 mm

TRACHEOBRONCHIAL REGION (TB)
lined with ciliated cells for rapid removal of particles (h)

RESPIRATORY BRONCHIOLE, 0.5 mm
ALVEOLAR DUCT, 0.2 mm
ALVEOLAR SAC, 0.3 mm
ALVEOLUS

PULMONARY REGION (P)
insoluble particles remain here for tens to hundreds of days

BRONCHIAL ARTERY
PULMONARY ARTERY
LYMPHATICS
PULMONARY VEIN
LYMPHATICS

Region of gas exchange oxygen uptake and carbon dioxide release

Fig. 7.1. Diagram of the respiratory tract.

There are about 5×10^8 alveoli, each on average about 150 μm in diameter. The walls of the NP and TB regions are covered with cilia, which propel the mucous layer out of the respiratory tract. Particles caught in the mucus are brought up and eventually swallowed.

The alveoli are not ciliated, and particles deposited in the P region remain there until they dissolve in lung fluid, or are engulfed by wandering cells called macrophages, which transport them to the ciliated bronchioles or to the lymphatic system of the lung.

7.2   Deposition of particles in the upper respiratory tract

Different mechanisms of deposition are effective in the three regions of the respiratory tract. In the NP and TB regions, flow velocities aɩe relatively high and transit times short, so impaction is of major importance, though diffusion and Brownian motion are also important for gases and very small particles. In the P region, velocities are very low, and the stay time of air in the lung during each breath allows sedimentation and Brownian movement to operate.

Hounam *et al.* (1971) measured deposition of radioactive particles in the NP region by drawing the aerosol in through the nose and out through the mouth of subjects who held their breath. Lippmann (1977), Stahlhofen *et al.* (1980) and Chan & Lippmann (1980) used collimated gamma scintillation counters to measure activity deposited in the head of subjects inhaling via the nose. A correction was made for activity removed by the mucociliary mechanism, swallowed and transferred to the stomach in the short interval between inhalation and measurement of the activity in the head. Yu *et al.* (1981) analysed these and other measurements of deposition in the NP region in terms of the parameter $\rho_p d_p^2 q$, where ρ_p and d_p are the density and diameter of the particles and q the flow rate of the inspired air. The parameter $\rho_p d_p^2$ is the square of the aerodynamic diameter of the particle (Section 6.4), and, assuming that the linear velocity in the airways is proportional to q, it is a function of the stopping distance of the particles. Figure 7.2 shows the correlation of the fractional deposition in the head with $\rho_p d_p^2 q$. A typical value of q in normal inspiration is 500 ml s^{-1}, and for a 10-μm diameter unit density particle this makes $\rho_p d_p^2 q$ equal to 5×10^4 g μm^2 s^{-1}. Figure 7.2 shows 90% deposition of 10-μm particles in the NP region with nose breathing, and 40% with mouth breathing, at this value of $\rho_p d_p^2 q$. If d_p is 1 μm, NP deposition is about 10% for nose breathing, and negligible for mouth breathing, at normal inspiratory flows.

7.3   Clearance of particles from the NP region

The Task Group on Lung Dynamics (1966) suggested a value of 4 min for the clearance half-life of particles in the NP region. To obtain experimental data, Fry & Black (1973) administered particles in the size range 2–10 μm, labelled with ^{99m}Tc, to subjects who inhaled the particles through the nose and exhaled through the mouth. Sodium iodide gamma ray detectors, collimated to respond to activity in the NP region, were used to monitor the removal of the particles. Particles deposited on the ciliated epithelium of the nose were rapidly removed to the pharynx and swallowed. The anterior (front) parts of the nasal passages are not ciliated, and particles deposited in this region were retained for periods up to several hours.

Clearance from the ciliated region may be impaired if the epithelium has been damaged. There has been an incidence of nasal adenocarcinoma in furniture workers exposed to wood dust. Black *et al.* (1974) inserted 5-μm polystyrene particles, labelled with ^{99m}Tc, onto the anterior end of the middle turbinate in the nose of furniture workers and control subjects. The clearance of the particles was monitored by measuring the gamma activity in the nose, using collimated sodium iodide detectors. In the control subjects, half the activity was removed in a mean time of 4.5 min (range 1.4 to 7.3 min), and it was calculated that the average rate of movement of the mucous film was 9 mm min^{-1}. The clearance was much slower in woodworkers who had been exposed to dust for 10 a or more. Certain types of wood contain soluble

Fig. 7.2. Statistical analysis of data on deposition of inhaled particles in the head (after Yu *et al.*, 1981).

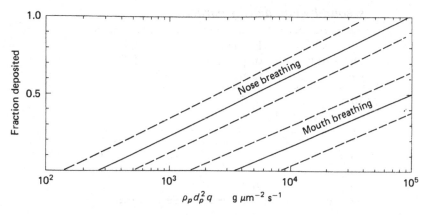

carcinogens, and, in workers with impaired clearance, the nasal epithelium is likely to be exposed to higher concentrations.

7.4   Deposition in the TB and P regions

Particles in the micrometre size range can be deposited by impaction in the TB region, particularly at the carina where the two main bronchi diverge. Using hollow casts of the trachea and main bronchi, Schlesinger *et al.* (1977) and Chan & Lippmann (1980) found that the efficiency of deposition correlated with the impaction parameter $\rho_p d_p^2 q$.

In the pulmonary region, air velocities are too low to impact particles small enough to reach that region, and the mechanisms of deposition are sedimentation and Brownian diffusion. The efficiency of both processes depends on the length of the respiratory cycle, which determines the stay time in the lung. If the cycle is 15 breaths/min, the stay time is of the order of a second. Table 7.1 shows the distance fallen in one second and the root mean square distance travelled by Brownian diffusion in one second by unit density particles (Fuchs, 1964). Sedimentation velocity is proportional to particle density, but Brownian motion is independent of density. Table 7.1 shows that sedimentation of unit density particles is more effective in causing deposition than Brownian diffusion when d_p exceeds $1\,\mu$m, whereas the reverse is true if d_p is less than $0.5\,\mu$m. For this reason, it is appropriate to use the aerodynamic diameter d_A equal to $\rho_p^{1/2} d_p$ when this exceeds $1\,\mu$m, but the actual diameter for submicrometre particles.

Table 7.1. *Root mean square Brownian displacement and distance fallen in one second by unit density particles*

Diameter (μm)	Brownian displacement (μm)	Distance fallen (μm)
10	1.7	3000
5	2.4	730
2	4.0	130
1	5.9	35
0.5	9.0	9.4
0.2	17	2.2
0.1	30	

It is difficult to distinguish particles in the TB and P regions by using collimated detectors, but deposition in the two regions can be separated by reference to the different rates of clearance from the two regions. Figure 7.3 shows typical graphs of the response of counters over the chests of subjects who have inhaled aerosols of 2-μm and 5-μm diameter particles. By curve stripping, the short and long components can be separated – though sometimes more than two components are found (Morrow & Yu, 1985). The intercepts of the dashed lines in Fig. 7.3 give the deposition in the P region as a fraction of that in the lung, that is P/(TB + P).

The fraction deposited in the TB and P regions varies between subjects, but in a single subject, P/(TB + P) varies systematically with particle size (Lippmann *et al.*, 1980; Stahlhofen *et al.*, 1980).

The first measurements of deposition and clearance were those of Albert & Arnett (1955), who used heterogeneous particles of iron oxide labelled with ^{59}Fe. In later work, to obtain monodisperse particles,

Fig. 7.3. Typical lung clearance curves: A, 2-μm particles, P/(TB + P) = 0.8; B, 5-μm particles, P/(TB + P) = 0.5.

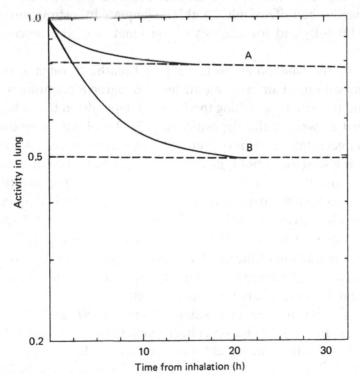

Albert *et al.* (1964, 1967) fed a dilute resuspension of iron oxide sol onto a spinning disk. In the outlet duct, the droplets were dried by infrared irradiation, leaving solid particles of size depending on the concentration of the original suspension. The Fe_2O_3 particles were labelled with ^{51}Cr or [198]Au, and no leaching of tracer was observed after storing the particles for several weeks in saline solution or in rat muscle. Because they were aggregates of primary sol particles with diameters of about 0.01 μm, the density of the particles inhaled was found to be 2.5 g cm^{-3}, which is less than that of solid Fe_2O_3. Numerous measurements of lung clearance using iron oxide particles, at the New York Institute of Environmental Medicine and elsewhere, have been summarised by Lippmann *et al.* (1980).

In the UK, Booker *et al.* (1967) used a spinning disk to make monodisperse polystyrene particles. Polystyrene was dissolved in xylene, at a concentration of 0.2%, and chromium acetyl acetonate, labelled with ^{51}Cr, was added. The spinning disk was operated at 3×10^4 rpm to produce 40-μm droplets of xylene which evaporated to give 5-μm polystyrene spheres. Few *et al.* (1970) adapted this method to produce particles of polystyrene labelled with [99m]Tc. This isotope has only very slight beta emission, so the dose to the lung is low, and though the radioactive half-life is only 6 h, this is adequate for estimation of the ratio P/(TB + P) and for analysis of the kinetics of the mucociliary clearance.

To inhale the labelled aerosols, subjects breathe through a mouthpiece. The volume of air in the mouth and TB region is respiratory dead space, and the activity reaching the P region depends on the volume of air inhaled as well as the deposition in TB. To obtain reproducible results, subjects take breaths of predetermined volume, and the breathing cycle is also standardised, usually at 14 or 15 breaths/min. Exhaled aerosol is collected on a filter, and the deposition in the mouth also estimated, so that the deposition in the lung (TB + P) is known as a fraction of the activity inhaled. The distribution as between TB and P is then determined by analysis of the clearance curves as in Fig. 7.3. Figure 7.4 shows the fraction of the inhaled activity deposited in the lung, and in the P region as a function of the particle size (aerodynamic diameter if greater than 0.5 μm, otherwise actual diameter).

The distinction in clearance kinetics from the NP and P regions, illustrated in Fig. 7.3, has however been challenged by Stahlhofen *et al.* (1986). These authors arranged for subjects to inhale Fe_2O_3 particles, labelled with [198]Au, in a bolus of air of defined volume. The bolus was introduced at various stages of the respiratory cycle in an attempt to label different parts of the tract with deposited particles. When a bolus of only 45 cm^3 volume was inhaled near the end of an inspired breath, the particles were expected to deposit only in the NP region. However, a lung term component in the clearance curve was found, implying either some penetration to the P region, or else slow clearance from some sites in the upper respiratory tract.

The lines in Fig. 7.4 are the results of theoretical calculations, using models of the respiratory tract (Yu & Diu, 1982). The points are measurements with radioactive aerosols. Numerous other determinations of fractional deposition in the whole tract have been made, using non-radioactive methods to count the number of particles in the inhaled and exhaled air (Heyder *et al.*, 1986; Schiller *et al.*, 1988). Fractional deposition is least for particles of about 0.2 to 0.5 μm diameter. Table 7.1 shows that the combined effect of sedimentation and Brownian motion is then at a minimum.

Hygroscopic particles grow as water vapour is condensed on them as they pass down the respiratory tract. Tu & Knutson (1989) found 50% deposition of 0.35-μm NaCl particles inhaled orally. Minimum deposition of NaCl was found at a particle size of 0.075 μm.

Fig. 7.4. Deposition of particles in alveolar region (open symbols) and whole respiratory tract (closed symbols) (14 or 15 breaths/min by mouth, tidal volume 1.0 to 1.5 l). Experimental results of Chan & Lippmann, 1980 ($\diamond$), Stahlhofen *et al.*, 1980 ($\square$), Foord *et al.*, 1978 ($\bigcirc$), Pritchard *et al.*, 1980 ($\triangle$). Error bars are 1 s.e. Lines are theoretical calculations of Yu & Diu, 1982.

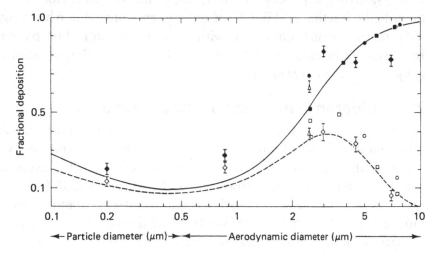

The results of Fig. 7.4 refer to oral breathing, with a ventilation rate of about 1 m^3 h^{-1}. If the ventilation rate is increased, by breathing more rapidly, more particles will be inhaled, but a smaller proportion will reach the pulmonary region because more will be lost by impaction in the upper respiratory tract. Deposition of particles of less than about 3 μm diameter is mainly determined by stay time in the lung. Larger particles are removed by impaction in the upper respiratory tract, and the proportion penetrating to the lung is reduced if the flow velocity is increased. Breathing through the nose instead of the mouth substantially reduces the fraction of particles in the micrometric range which penetrate to the lung, but there is little difference with submicrometric particles (Tu & Knutson, 1984), as can be deduced from Fig. 7.2.

7.5   Effect of smoking and of disease on mucociliary clearance

If mucociliary clearance were impaired by heavy smoking, carcinogens deposited in the bronchial passages would be retained for longer periods, and this might be a factor in the aetiology of lung cancer. However, tests on subjects who smoke have shown little effect on the clearance during the initial rapid phase. Pavia *et al.* (1970) measured clearance of 5-μm polystyrene particles, labelled with ^{99m}Tc, from the lungs of elderly subjects, and found no difference in results between life-long heavy smokers, non-smokers and ex-smokers. Thomson & Pavia (1974) compared clearance of 5-μm particles in healthy smokers and bronchitic smokers, and found deeper penetration in the healthy lungs, probably because air velocities were higher in the obstructed TB airways, giving more deposition by impaction in this region. Consequently the fraction P/(TB + P) of activity remaining after 6 h averaged 70% in the normal compared with 53% in the bronchitic patients. Increased deposition has also been seen in the TB region of asthmatic subjects (Lippmann *et al.*, 1971).

7.6   Clearance from the pulmonary region

To calculate the dose to the lung from inhalation of insoluble particulate activity of long radioactive half-life, it is necessary to estimate the clearance from the P region. ICRP (1979) assumed a clearance half-life of 500 d. Booker *et al.* (1967) found half-lives in the range 150–300 d in subjects who had inhaled polystyrene particles labelled with ^{51}Cr, but the radioactive half-life of this isotope (27 d) restricted the duration of the measurements to 100 d. In more recent work, Bohning *et al.* (1982) used 3.6-μm diameter polystyrene particles labelled with [85]Sr (half-life 65 d). Bailey *et al.* (1982, 1985) used [85]Sr to label 1.2-μm fused aluminosilicate (FAP) particles and [88]Y (half-life 108 d) with 3.9-μm FAP. Bailey *et al.*'s subjects inhaled the 1.2- and 3.9-μm particles in the same day.

The results of Bohning *et al.*'s and Bailey *et al.*'s experiments were similar. The clearance from the P region of non-smokers and ex-smokers was fitted by double exponential curves. A fraction of the activity was cleared with a half-life of 20–30 d, this percentage being larger for the 3.9-μm than for the 1.2-μm particles. Of the remaining activity, half was cleared in about a year. Cigarette smoking suppressed the fast pulmonary clearance phase, and increased the half-life of the slow phase. Persons with obstructive lung disease also showed increased half-life of the slow phase. Bailey *et al.* measured the activity of [85]Sr and [88]Y in the urine of their subjects, and showed that there was some systemic clearance. They inferred that the observed lung clearance was not entirely due to mechanical removal of particles, dissolution also contributing. They estimated that the half-time for mechanical removal might be as long as 600 d.

7.7 Cigarette tar

Mainstream cigarette smoke contains about 10^{10} submicrometre particles per millilitre. These particles constitute the solid matter known as tar. For animal experiments, the hexadecane in tar can be labelled with ^{14}C, but this method is unsuitable for studies with humans, as movement of activity in the living subject cannot be followed unless there is gamma emission. At Harwell, Black & Pritchard (1984) used 1-iodo-hexadecane (1-IHD) labelled with [123]I. This compound is non-toxic, and its boiling point (380°C) lies in the range of boiling points of the constituents of tar. The reaction between 1-hexadecane with phosphorus tri-iodide was used to synthesise 1-IHD, and labelling with ^{123}I was achieved by an exchange reaction in a solution of 10% water in acetone. Following the exchange, methylene chloride was added, and the solution shaken with aqueous sodium thiosulphate to remove free iodine. Labelled 1-IHD was separated from the organic phase, and was introduced into the cigarette with a syringe pump. To check that the tracer was fixed to the tar, chemical tests and gas chromatography were done on particles collected from smoke of labelled cigarettes. These showed that over 99% of the ^{123}I was associated with 1-IHD. The activity median aerodynamic diameter of the smoke particles as estimated from the catch of ^{123}I on the stages of a cascade impactor was 0.8 μm.

Volunteers smoked the labelled cigarettes. The activity deposited in the mouth, and the activity removed to the GI tract in the 15-min interval between smoking and measurement, were measured with collimated detectors. The clearance of activity from the lung was followed for 2 d to allow the deposit to be fractionated into TB and P components. Comparative tests were also done with subjects inhaling 2.5-μm polystyrene particles, which were also labelled with ^{123}I.

Table 7.2 shows the percentage deposition of polystyrene particles and tar in the mouth, TB and P compartments (Pritchard & Black, 1984). The distribution of polystyrene particles was as expected from other work, but the distribution of tar appeared inconsistent with the submicrometre particle size. The ratio P/(TB + P) was 0.37, a value which is typical of an aerosol with diameter 6 μm. It is believed that the tobacco smoke particles grew hygroscopically in the upper respiratory tract. Also, coagulation may have occurred in the mouth region. Whatever the explanation, the effective deposition of cigarette tar in the TB region may be significant in the aetiology of lung cancer.

The polystyrene particles and the 1-IHD labelled tar were removed from the TB region with similar clearance half-lives of about 2 h (Fig. 7.5). The pulmonary removal of polystyrene was too slow to measure with the ^{123}I label, but 1-IHD was removed from the P region with a half-life of 18 h, and this was attributed to absorption into the bloodstream.

Table 7.2. *Regional deposition in the lung*

Aerosol	No. of tests	Percentage of deposit (mean ± S.E.)			P/(TB + P)
		Mouth	TB	P	
2.5-μm polystyrene	52	15 ± 3	25 ± 4	60 ± 3	0.71
Cigarette tar	97	17 ± 2	52 ± 2	31 ± 3	0.37

7.8   Airborne lead from vehicle exhausts

Tetra-ethyl lead is added to petrol as an anti-knock agent. When the petrol is burnt in the engine, the organic lead is converted to inorganic compounds. Ethylene dibromide is added to the tetra-ethyl lead as a scavenger, to limit the build-up of lead in the engine. Lead is emitted from the exhaust as a very fine aerosol, consisting mainly of lead halide, PbBrCl (Piver, 1977). In the atmosphere, chemical reactions yield a variety of compounds. Biggins & Harrison (1978), using X-ray powder diffraction on filter samples, identified $(NH_4)_2SO_4.PbSO_4$ together with lesser amounts of $PbSO_4$ and PbBrCl. They considered that $(NH_4)_2SO_4.PbSO_4$ was an intermediate in the conversion of halides to $PbSO_4$, formed by reaction with atmospheric ammonia and acid sulphates.

The primary lead particles produced by an engine running at high speed are typically 0.02 to 0.05 μm in diameter. Accumulation and coagulation cause the particle size to increase with time in the atmosphere. Figure 7.6 shows electromicrographs of particles collected by thermal precipitation at the side of a motorway (Little & Wiffen, 1978). Some of the lead particles are discrete, others are attached to carbonaceous chain aggregates, which are diesel smoke. Carbonaceous chain aggregates are also produced by petrol engines when idling. Measure-

Fig. 7.5. Lung clearance curves of one subject: ○, polystyrene particles, 2.5-μm diameter; △, 1-iodohexadecane in cigarette tar.

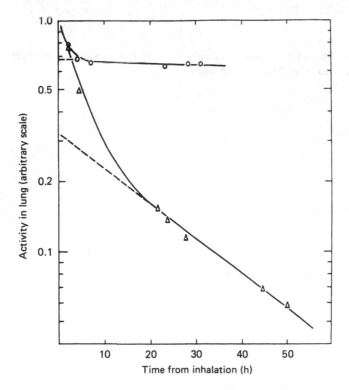

ments with cascade impactors near busy roads have shown that more than 50% of the lead may be associated with particles of diameter less than 0.3 μm (Chamberlain *et al.*, 1978).

7.9   Inhalation of lead aerosols

As early as 1937, R.A. Kehoe began to investigate the human uptake of lead at the Kettering Laboratory, Cincinnati. A full account of the work, with statistical analysis, has been published by Gross (1981). In this and later work by Griffin *et al.* (1975), lead aerosol was produced by burning tetra-ethyl lead in propane and was passed into chambers. Volunteers were exposed in the chambers to the lead aerosol daily over periods of several months. The concentration of lead in the air (PbA) was monitored continuously, and samples of venous blood were taken from the volunteers at intervals for measurement of blood lead (PbB). It was found that PbB increased during the first month or two and then reached a quasi-equilibrium in which the intake from inhalation was balanced by excretion.

The response of blood lead to air lead can be expressed as

$$\alpha = \frac{\text{increase in PbB}}{\text{PbA}}$$

Fig. 7.6. A typical motorway vehicle exhaust aerosol viewed by electron microscopy at three magnifications.

0.1 μm   1.0 μm   4.0 μm

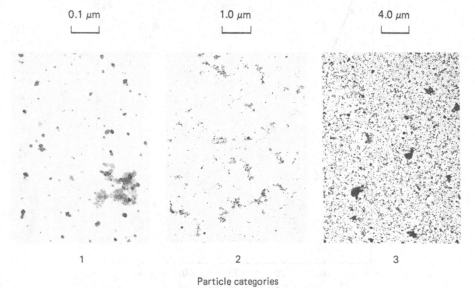

1   2   3

Particle categories

If PbB is measured in μg kg^{-1} and PbA in μg m^{-3}, α has units m^3 kg^{-1}. In cases where the volunteer was exposed for a certain number of hours per day, PbA is taken as the equivalent 24-h average. In principle, estimates of α can also be made by comparing PbB in populations exposed to differing PbA in the ambient air, on the assumption, rarely fulfilled in practice, that uptakes of lead from dietary and other sources are the same. As the uptake of lead from all sources increases, PbB is found to increase non-linearly (references in Chamberlain, 1985), so the hypothetical relation between PbB and PbA, with a constant level of dietary lead, is as shown in Fig. 7.7. The tangent AP has slope α, and α is expected to decrease as PbB increases.

During the exposure periods, the PbB of Kehoe's and Griffin et al.'s subjects was in the range 200–500 μg kg^{-1}, high by current standards. The values of α derived from their experiments range from 4 to 17 m^3 kg^{-1}, with some evidence of an inverse relation with PbB (Chamberlain, 1983). It is also possible that higher values of α would have been obtained if the experiments had continued for years rather than months.

Fig. 7.7. Hypothetical relation between lead in blood (PbB) and lead in air (PbA).

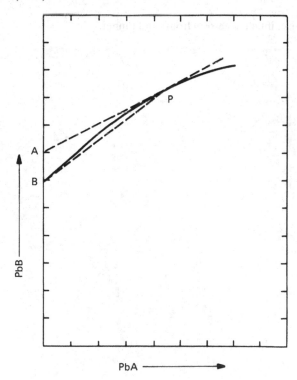

An advantage of the use of radioactive tracers is that uptake and elimination of a potentially toxic substance can be studied without any perceptible increase in its concentration in the body. Also, the criticism was made that the aerosols used by Kehoe and by Griffin *et al.* were not identical with those produced in motor exhaust. To study uptake following inhalation of exhaust lead, a method of incorporating a tracer was developed at Harwell.

7.10 Radioactive labelling of exhaust lead

The isotope ^{203}Pb is produced by bombardment of thallium with protons in a cyclotron. The radioactive half-life is 52 h and decay is by electron capture. The associated 279-keV gamma ray enables the activity in the lung and other organs of volunteer subjects to be detected by external gamma ray spectrometers (Chamberlain *et al.*, 1975).

^{203}Pb was obtained as a solution of $PbCl_2$. Tetra-ethyl lead was synthesised by a Grignard reaction and added to petrol together with ethylene dibromide. The petrol was burnt in a 50-cc four-stroke engine, and the exhaust passed into a wind tunnel (Fig. 7.8). The airflow in the tunnel gave sufficient dilution to limit the concentration of CO to 1000

Fig. 7.8. Subject inhaling aerosol from wind tunnel.

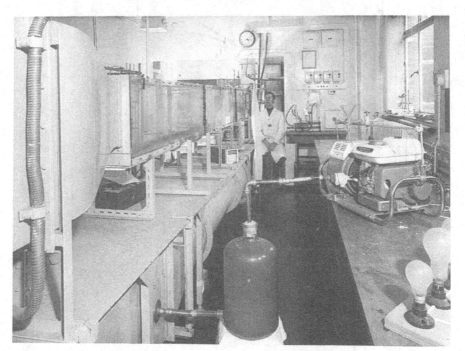

ppm. Lead aerosols with mean diameters 0.02, 0.04 and 0.09 μm were obtained by varying the stay time in the system between the engine and the wind tunnel. The particle size was estimated visually under the microscope and also by the use of a diffusion battery to measure the Brownian diffusivity of the particle.

Experiments were also done with an aggregated lead aerosol, with particle size about 0.5 μm. This was obtained by storing the exhaust aerosol in a 600-l chamber for about 50 min. The floor of the chamber was covered with an oxidation catalyst to remove CO. By varying the performance of the engine an aerosol of chain-aggregated carbon-aceous particles, with lead attached, was obtained. Also, to investigate the effect of varying chemical composition, lead nitrate and lead oxide particles were produced by burning tetra-ethyl lead in propane with and without nitrogen.

Subjects inhaled from the wind tunnel or the chamber and exhaled through a filter. The activity of [203]Pb deposited in the lung was estimated from the activity inhaled and exhaled and also from the response of counters above and below the chest (Fig. 7.9). The inhalation took a few minutes, and the first measurement of activity in the lung was made immediately. As [203]Pb passed from the lung to the blood-stream, the count from the activity in the lung decreased, but the count from other activity in the chest increased. To measure the loss from the

Fig. 7.9. Measurement of [203]Pb in chest.

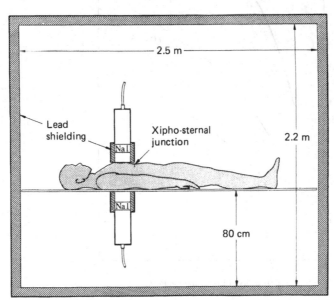

lung, a correction was made. This was based on the activity in blood, as measured in sequential samples of venous blood. Additional calibration experiments in which each subject was given ^{203}Pb by saline injection, were done to relate the counter response to the systemic activity.

7.11 Deposition in the lungs of ^{203}Pb-labelled exhaust particles

Figure 7.10 shows the percentage deposition of the exhaust aerosol in the lung as a function of the length of the respiratory cycle (Chamberlain *et al.*, 1978). Eight volunteers inhaled the aerosols. Tidal volumes ranged from 0.3 to 2.3 l. Individual results showed a coefficient of variation of 20% relative to the curves of Fig. 7.10. In the experiments with the wind-tunnel aerosols (0.02, 0.04 and 0.09 μm) there was no significant effect of tidal volume on the percentage deposition.

Fig. 7.10. Deposition in lung of ^{203}Pb-labelled particles.

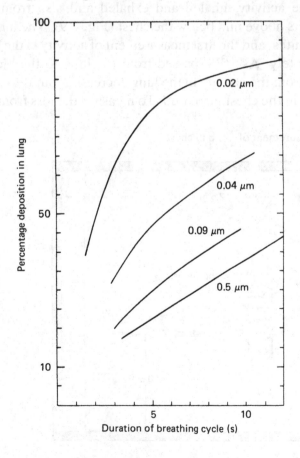

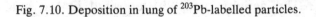

Deposition of the aggregated aerosol (0.5 μm) increased with increasing tidal volume (Wells *et al.*, 1977). The lines in Fig. 7.10 represent results normalised to a tidal volume of 1.0 l.

A comparison of the results with other data on the deposition of submicrometre aerosols, all related to a tidal volume of 1 l, is shown in Fig. 1.14. Although the density of the lead particles was greater than that of the other particles, the fractional deposition was similar, except possibly for the 0.5 μm size, because deposition was by Brownian motion. The percentage deposition increases for particles of diameter less than 0.1 μm, and this means increased uptake of lead, relative to a given PbA, for persons exposed to non-aggregated aerosol, as found alongside major roads.

7.12 Clearance of ^{203}Pb from the lung and transfer to blood

Table 7.3 shows the percentage of the original burden of lead in the lung remaining at 24 h after inhalation. Comparison of the results for lead nitrate and lead oxide show some effect of solubility, but this is slight. Particles of submicrometre size dissolve in lung fluids even though the solubility of the compound in bulk may be low. Morrow *et al.* (1980) found no difference in the rate of clearance of submicrometre aerosols of $PbCl_2$ and $Pb(OH)_2$ from the lungs of volunteer subjects. Both Chamberlain *et al.* (1978) and Morrow *et al.* varied the amount of stable lead carrier in the aerosol, and this also made no difference to the rate of clearance from the lung.

From the activity in samples of venous blood, and the subject's estimated blood volume, the transfer of ^{203}Pb from lung to blood could

Table 7.3. *Percentage of ^{203}Pb-labelled particles remaining in lung 24 h after inhalation*

Aerosol	Particle size (μm)	Percentage remaining at 24 h
Exhaust	0.02	13
Exhaust	0.5	11
Carbonaceous exhaust	0.5	18
Lead oxide	0.75	15
Lead nitrate	0.75	8

be derived. Figure 7.11 shows results obtained with [203]Pb-labelled exhaust (Chamberlain *et al.*, 1975), and also with [212]Pb attached to condensation nuclei (Hursh *et al.*, 1969; Booker *et al.*, 1969). In all three series of experiments, the concentration in blood reached a peak about 30 h after the inhalation, and was equivalent to 50–60% transfer of lead from lung to blood. This does not mean that nearly half the lead remained in the lung, because a similar percentage of the initial dose was found in blood when the activity was given by injection as a saline solution. When tracer amounts of lead enter blood plasma, some exchanges into extracellular fluid or becomes immobilised at storage sites in various organs. The activity of [203]Pb remaining in blood, after the equilibration process was complete, was found to be attached to red cells, less than 1% being in plasma (Chamberlain *et al.*, 1975).

Measurements of activity in blood of seven subjects were continued for 14 d after inhalation and showed that lead was lost from blood with a biological half-life of $18.0 \pm$ (s.e.) 0.9 d. Because of radioactive decay, it was not possible to continue measurements for a longer period. Rabinowitz *et al.* (1976) gave oral doses of the stable lead isotope [204]Pb to four subjects daily for several months, and used mass spectrometry to measure [204]Pb in blood. The decline in [204]Pb in blood was found to be exponential over periods of about 100 d from the end of the period of

Fig. 7.11. Uptake of radiolead from lung to blood: ×, Booker *et al.*, 1969; ○, Hursh *et al.*, 1969; △, Chamberlain *et al.*, 1975.

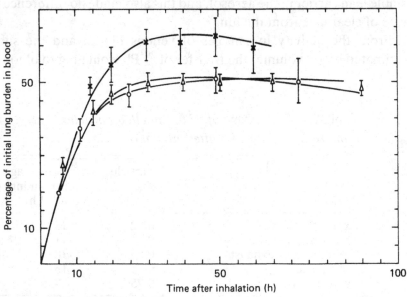

dosing. The biological half-lives were 18, 24, 36 and 18 d for the four subjects.

Over periods of years, resorption of lead from storage in bone gives a longer apparent biological half-life to lead in blood. No tracer experiments with lead have continued for a long enough time to evaluate the entry into blood of resorbed lead. On the assumption that the rates of resorption of lead are similar to those found with ^{90}Sr and other alkaline earth elements, compartmental analysis suggests that resorbed lead may contribute 40% of the input to blood when exposure has lasted for 50 a (Chamberlain, 1985).

7.13   Contribution of petrol lead to blood lead

The results of the volunteer experiments with ^{203}Pb and ^{204}Pb can be used to derive a value for the factor β defined as the contribution of PbA to PbB. In Fig. 7.7, β is the slope of the line BP, and is therefore not quite the same as α, which is the slope of AP. The following assumptions are made:

(i)   Ventilation rate of the average adult is 15 m^3 d^{-1}. From physiological data (Cotes, 1965; ICRP, 1975) it can be shown that this corresponds to the standard daily dietary intake of 3000 kcal.

(ii)   50% of the inhaled aerosol is deposited in the lung and absorbed.

(iii)   55% of the absorbed lead becomes attached to red cells.

(iv)   Biological half-life in blood is 21 d (mean of results with ^{203}Pb and ^{204}Pb), corresponding to a mean life of 30 d.

(v)   Mass of blood in standard man is 5.4 kg (ICRP, 1975).

With these assumptions

$$\beta = \frac{15 \times 0.5 \times 0.55 \times 30}{5.4} = 23 \,\mu g \, kg^{-1} \, per \, \mu g \, m^{-3}$$

This value should be increased, possibly to about 30 m^3 kg^{-1}, in the long term to allow for resorption of lead from bone (Chamberlain, 1985).

A comparison can be made with the results of the Isotopic Lead Experiment carried out in Piedmont, Italy (Facchetti & Geiss, 1982). During three years, 1977–9, the tetra-ethyl lead used in Piedmont was manufactured using exclusively Australian lead, which has a low content of the stable isotope ^{206}Pb. As a result, the ratio ^{206}Pb/^{207}Pb in petrol was changed from 1.18 to 1.04. In Turin, the average ratio

^{206}Pb/^{207}Pb in adult blood decreased from 1.628 to 1.325. Assuming that this change was due entirely to the change in petrol lead, 25% of the lead in blood must have derived from petrol, either by inhalation or by some other route.

Turin has a high density of motor traffic, and at the time of the Isotope Lead Experiment, Italian petrol had 0.6 g l^{-1} of lead. Windspeeds in Piedmont are low, due to shelter by the Alps. Average PbA in residential areas of Turin was about 1.7 μg m^{-3} (Facchetti & Geiss, 1982). The mean PbB of adults was 236 μg kg^{-1}. Hence

$$\beta = \frac{236 \times 0.25}{1.7} = 35 \text{ m}^3 \text{ kg}^{-1}$$

The experiment was not of sufficient duration to attain complete equilibrium between PbB and PbA, including transfer to and from storage in bone. However, there was also some transfer of lead to diet, through fallout on crops.

In January 1986, the permissible concentration of lead in petrol in the UK was reduced by 62%, from 0.4 to 0.15 g l^{-1}. Regular measurements of PbA in 1985 and 1986 have been reported from 26 urban locations (McInnes, 1986; Jensen & Laxen, 1987; Pattenden & Branson, 1987; Page *et al.*, 1988). The mean percentage fall in PbA was 49.4 $\pm$ (s.e.) 1.9%. This is in accordance with other evidence that before 1986 lead from petrol contributed about 80% of airborne lead.

The Department of the Environment carried out surveys of lead in blood over the four years 1984 to 1987 (Quinn & Delves, 1987; Department of the Environment, 1988). Cohorts of adult subjects were followed, with replacements where necessary. The exposed group were residents of houses along busy roads. A separate group of traffic police was included. Schools adjacent to busy roads were visited and blood was taken from children in the same class each year. All groups showed falls in PbB as between 1984/5 and 1986/7, as well as between 1985/6, when the lead content of petrol was reduced. The percentage falls in 1984/5 and 1986/7 were not significantly different, and averaged 6.2%. Comparison with earlier surveys of blood lead indicated a continuing falling trend at least since 1979. This is ascribed to reductions in dietary and other sources of lead.

In Table 7.4, the percentage reductions in PbB in 1985/6, coincident with the reduction in lead in petrol, for the exposed groups are shown. Assuming 6.2% reduction from other causes, the percentage reductions associated with the change in petrol, are as shown in the second column.

Table 7.4. *Reductions in PbB from D.O.E. survey*

	Reduction in 1985/6 (%)	Excess over trend (%)	S.E. of excess (%)	Excess over trend (μg kg^{-1})
Exposed men and women	10.3	4.1	1.5	4.2
Traffic police	18.9	12.7	2.7	13.7
Children	15.8	9.6	1.5	8.7

The third column gives an approximate estimate of the standard error of the excess reduction, taking into account only statistical variations. The final column of Table 7.4 gives the excess reductions in terms of PbB. Since a reduction in petrol lead of 60% gave a reduction in PbB of exposed men and women of 4%, the conclusion might be drawn that lead in petrol contributed 7% of PbB before the reduction. Similarly, for traffic police and children, it contributed 20% and 16% respectively. These results are broadly compatible with the results of the Isotopic Lead Experiment, taking into account the special factors tending to increase exposure to PbA in Turin.

The 26 stations mentioned above include both roadside and urban background locations, and gave mean PbA levels of 0.70 before and 0.35 μg m^{-3} after the change in petrol. Assuming that a reduction of 0.35 μg m^{-3} resulted in the falls in PbA of the last column of Table 7.4, β can be estimated as 7, 39 and 25 m^3 kg^{-1} for exposed men and women, traffic police, and children respectively. Thus the falls in PbB in these subjects, coincident with the reduction in lead in petrol, are reasonably consistent with expectations based on the inhalation experiments with ^{203}Pb and the Isotopic Lead Experiment in Piedmont.

References

Albert, R.E. & Arnett, L.C. (1955) Clearance of radioactive dust from the lung. *American Medical Association Archives of Industrial Health*, **12**, 99–106.

Albert, R.E., Lippmann, M., Spiegelman, J., Strehlow, C., Briscoe, W., Wolfson, P. & Nelson, N. (1967) The clearance of radioactive particles from the human lung. In: *Inhaled Particles and Vapours*, *II*., ed. C.N. Davies, pp. 361–77. Oxford: Pergamon.

Albert, R.E., Petrow, H.G., Salam, A.S. & Spiegelman, J.R. (1964) Fabrication of monodisperse lucite and iron oxide particles with a spinning disk generator. *Health Physics*, **10**, 933.

Bailey, M.R., Fry, F.A. & James, A.C. (1982). The long term clearance kinetics of insoluble particles from the human lung. *Annals of Occupational Hygiene*, **26**, 273–90.

(1985) Long term retention of particles in the human respiratory tract. *Journal of Aerosol Science*, **16**, 295–305.

Biggins, P.D.E. & Harrison, R.M. (1978) Identification of Pb compounds in urban air. *Nature*, **272**, 531–2.

Black, A. & Pritchard, J.N. (1984) A comparison of the regional deposition and short term clearance of tar particulate matter from cigarette smoke with that of 2.5 μm polystyrene microspheres. *Journal of Aerosol Science*, **15**, 224–7.

Black, A., Evans, J.C., Hadfield, E.H., Macbeth, R.G., Morgan, A. & Walsh, M. (1974) Impairment of nasal mucociliary clearance in woodworkers in the furniture industry. *British Journal of Industrial Medicine*, **31**, 10–17.

Bohning, D.E., Atkins, H.,L. & Cohn, S.H. (1982) Long term particle clearance in man: normal and impaired. *Annals of Occupational Hygiene*, **26**, 259–71.

Booker, D.V., Chamberlain, A.C., Rundo, J., Muir, D.C.F. & Thomson, M.L. (1967) Elimination of 5 μ particles from the human lung. *Nature*, **215**, 30–3.

Booker, D.V., Chamberlain, A.C., Newton, D. & Stott, A.N.B. (1969) Uptake of radioactive lead following inhalation and injection. *British Journal of Radiology*, **42**, 457–66.

Chamberlain, A.C. (1983) Effect of airborne lead on blood lead. *Atmospheric Environment*, **17**, 677–92.

(1985) Prediction of response of blood lead to airborne and dietary lead from volunteer experiments. *Proceedings of the Royal Society B*, **224**, 149–82.

Chamberlain, A.C., Clough, W.S., Heard, M.J., Newton, D., Stott, A.N.B. & Wells, A.C. (1975) Uptake of lead by inhalation of motor exhaust. *Proceedings of the Royal Society B*, **192**, 77–110.

Chamberlain, A.C., Heard, M.J., Little, P., Newton, D., Wells, A.C. & Wiffen, R.D. (1978). Investigations into lead from motor vehicles. UKAEA Report AERE – R 9198. London, HMSO.

Chan, T.L. & Lippmann, M. (1980) Experimental measurements and empirical modelling of the regional deposition of inhaled particles in humans. *American Industrial Hygiene Journal*, **41**, 399–409.

Cotes, J.E. (1965) *Lung Function*. London: Blackwell.

Department of The Environment (1988) Digest of environmental protection and water statistics, No. 11. London, HMSO.

Facchetti, S. & Geiss, F. (1982) Isotopic lead experiment – status report. Luxembourg: CEC.

Few, J.D., Short, M.D. & Thomson, M.L. (1970) Preparation of ^{99m}Tc labelled particles for aerosol studies. *Radiochemical Radioanalytical Letters*, **5**, 275–7.

Foord, N., Black, A. & Walsh, M. (1978) Regional deposition of 2.5–7.5 μm diameter particles in healthy male non-smokers. *Journal of Aerosol Science*, **9**, 343–57.

Fry, F.A. & Black, A. (1973) Regional deposition and clearance of particles in the human nose. *Journal of Aerosol Science*, **4**, 113–24.

Fuchs, N.A. (1964) *The Mechanics of Aerosols*. Oxford: Pergamon.

Griffin, T.B., Coulston, F., Wells, H., Russell, J.C. & Knelson, J.H. (1975) Clinical studies on men continuously exposed to airborne particulate lead. In: *Lead*, ed. T.B. Griffin & J.H. Knelson, pp. 221–40. New York: Academic Press.

Gross, S.B. (1981) Human oral and inhalation exposures to lead: summary of the Kehoe balance experiments. *Journal of Toxicology & Environmental Health*, **8**, 333–71.

Heyder, J., Gebhart, J., Rudolf, G., Schiller, C.F. & Stahlhofen, W. (1986) Deposition of particles in the human respiratory tract in the size range 0.005 μm–15 μm. *Journal of Aerosol Science*, **17**, 811–25.

Hounam, R.F., Black, A. & Walsh, M. (1971) The deposition of aerosol particles in the nasopharyngeal region of the human respiratory tract. *Journal of Aerosol Science*, **2**, 47–61.

Hursh, J.B., Schraub, A., Sattler, E.L. & Hofmann, H.P. (1969) Fate of ^{212}Pb inhaled by human subjects. *Health Physics*, **16**, 257–67.

International Commission on Radiological Protection (1975) Report of the Task Group on Reference Man. Publication No. 23. Oxford: Pergamon.

(1979) Limits for intake of radionuclides by workers. ICRP Publication 30. *Annals of the ICRP*, **2**, (3/4).

Jensen, R.A. & Laxen, D.P.H. (1957) The effect of the phase-down of lead in petrol on levels of lead in air. *The Science of the Total Environment*, **59**, 1–8.

Lippmann, M. (1977) Regional deposition of particles in the human respiratory tract. In: *Handbook of Physiology – Reaction to Environmental Agents*, ed. D.H.K. Lee, pp. 213–32. Bethesda, Md.: The American Physiological Society.

Lippmann, M., Albert, R.E. & Paterson, H.T. (1971) The regional deposition of inhaled aerosols in man. In: *Inhaled Particles*, *III*, vol. 1, ed. W.H. Walton, pp. 105–210. Old Woking, Surrey: Gresham Press.

Lippmann, M., Yeates, D.B. & Albert, R.E. (1980) Deposition, retention and clearance of inhaled particles. *British Journal of Industrial Medicine*, **37**, 337–62.

Little, P. & Wiffen, R.D. (1978) Emission and deposition of lead from motor exhausts. II Airborne concentration, particle size and deposition of lead near motorways. *Atmospheric Environment*, **12**, 1331–41.

McInnes, G. (1986) Airborne lead concentrations and the effect of reductions in the lead content of petrol. Report LR 587(AP)M. Stevenage: Warren Spring Laboratory.

Morrow, P.E., Beiter, H., Amato, F. & Gibb, F.R. (1980) Pulmonary retention of lead: an experimental study in man. *Environmental Research*, **21**, 373–84.

Morrow, P.E. & Yu, C.P. (1985) Models of aerosol behaviour in airways. In: *Aerosols in Medicine, Principles, Diagnosis and Therapy*, ed. M.T. Newhouse & M.B. Dolovich. Amsterdam: Elsevier.

Page, R.A., Cawse, P.A. & Baker, S.J. (1988) The effect of reducing petrol lead on airborne lead in Wales, U.K. *The Science of the Total Environment*, **68**, 71–7.

Pattenden, N.J. & Branson, J.R. (1987) Relation between lead in air and in petrol in two urban areas of Britain. *Atmospheric Environment*, **21**, 2481–3.

Pavia, D., Short, M.D. & Thomson, M.L. (1970) No demonstrable long term effects of cigarette smoking on the mucociliary mechanism of the human lung. *Nature*, **226**, 1228–31.

Piver, W.T. (1977) Environmental transport and transformation of automotive-emitted lead. *Environmental Health Perspectives*, **19**, 247–59.

Pritchard, J.N. *et al.* (1980) A comparison of the regional deposition of monodisperse polystyrene particles in the respiratory tract of healthy male smokers and non-smokers. In: *Aerosols in Science, Medicine and Technology*, ed. W. Stöber & D. Hochrainer. Schmallenberg, F.R.G.: Gesellschaft für Aerosolforschung.

Pritchard, J.N. & Black, A. (1984) An estimate of the tar particulate matter depositing in the respiratory tracts of healthy male middle- and low-tar cigarette smokers. In: *Aerosols*, ed. B.Y.H. Liu & D.Y.H. Piu. Amsterdam: Elsevier.

Quinn, M.J. & Delves, H.T. (1987) UK blood monitoring programme 1984–1987: protocol and results for 1984. *Human Toxicology*, **6**, 459–74.

Rabinowitz, M.B., Wetherill, G.W. & Kopple, J.D. (1976) Kinetic analysis of lead metabolism in healthy humans. *Journal of Clinical Investigation*, **58**, 260–70.

Schiller, C.F., Gebhart, J., Heyder, J., Rudolf, G. & Stahlhofen, W. (1988) Deposition of monodisperse insoluble aerosol particles in the 0.005 to 0.2 μm size range within the human respiratory tract. *Annals of Occupational Hygiene*, **32**, 41–9.

Schlesinger, R.B., Bohning, D.E., Chan, T.L. & Lippmann, M. (1977) Particle deposition in a hollow cast of the human tracheobronchial tree. *Journal of Aerosol Science*, **8**, 429–41.

Stahlhofen, W., Gebhart, J. & Heyder, J. (1980) Experimental determination of regional deposition of aerosol particles in the human respiratory tract. *American Industrial Hygiene Journal*, **41**, 385–98.

Stahlhofen, W., Gebhart, J., Rudolf, G. & Scheuck, G. (1986) Measurement of lung clearance with pulses of radioactively-labelled aerosols. *Journal of Aerosol Science*, **17**, 333–6.

Task Group on Lung Dynamics (1966) Deposition and retention models for internal dosimetry of the human respiratory tract. *Health Physics*, **12**, 173–267.

Thomson, M.L. & Pavia, D. (1974) Particle penetration and clearance in the human lung. *Archives of Environmental Health*, **29**, 214–19.

Tu, K.W. & Knutson, E.O. (1984) Total deposition of ultrafine hydrophobic and hygroscopic aerosols in the human respiratory tract. *Aerosol Science and Technology*, **3**, 453–65.

Wells, A.C., Venn, J.B. & Heard, M.J. (1977) Deposition of the lung and uptake to blood of motor exhaust labelled with [203]Pb. In: *Inhaled Particles*, *IV*, ed. W.H. Walton, pp. 175–88. Oxford: Pergamon.

Yu, C.P. & Diu, C.K. (1982) A comparative study of aerosol deposition in different lung models. *American Industrial Hygiene Journal*, **43**, 54–65.

Yu, C.P., Diu, C.K. & Soong, T.T. (1981) Statistical analysis of aerosol deposition in the nose and mouth. *American Industrial Hygiene Association Journal*, **42**, 726–33.

INDEX

Italics show page where definition is given; bold type, principal references.

Printed in the United States
By Bookmasters